Leveraging Commercial Space Services

Opportunities and Risks for the Department of the Air Force

JONATHAN P. WONG, YOOL KIM, KRISTA LANGELAND, GEORGE NACOUZI,
KRISTA ROMITA GROCHOLSKI, JONATHAN BALK, KARISHMA V. PATEL,
BARBARA BICKSLER

Prepared for the Department of the Air Force
Approved for public release; distribution is unlimited

RAND | PROJECT AIR FORCE

For more information on this publication, visit **www.rand.org/t/RRA1724-1**.

About RAND

The RAND Corporation is a research organization that develops solutions to public policy challenges to help make communities throughout the world safer and more secure, healthier and more prosperous. RAND is nonprofit, nonpartisan, and committed to the public interest. To learn more about RAND, visit www.rand.org.

Research Integrity

Our mission to help improve policy and decisionmaking through research and analysis is enabled through our core values of quality and objectivity and our unwavering commitment to the highest level of integrity and ethical behavior. To help ensure our research and analysis are rigorous, objective, and nonpartisan, we subject our research publications to a robust and exacting quality-assurance process; avoid both the appearance and reality of financial and other conflicts of interest through staff training, project screening, and a policy of mandatory disclosure; and pursue transparency in our research engagements through our commitment to the open publication of our research findings and recommendations, disclosure of the source of funding of published research, and policies to ensure intellectual independence. For more information, visit www.rand.org/about/research-integrity.

RAND's publications do not necessarily reflect the opinions of its research clients and sponsors.

Published by the RAND Corporation, Santa Monica, Calif.
© 2023 RAND Corporation
RAND® is a registered trademark.

Library of Congress Cataloging-in-Publication Data is available for this publication.

ISBN: 978-1-9774-1129-7

Cover: U.S. Defense Advanced Research Projects Agency.

Limited Print and Electronic Distribution Rights

About This Report

The expanding global commercial space industry offers new opportunities for the U.S. Department of Defense (DoD). Technological advances in small satellites, lower launch costs, and innovative satellite applications may help DoD meet its mission requirements more efficiently and give it access to a broader innovation pool. The Department of the Air Force (DAF), particularly the U.S. Space Force (USSF), is increasing its partnership activities with the commercial space industry to take advantage of these opportunities. As the DAF makes investment decisions to leverage commercial space capabilities, it needs a better understanding of opportunities, risks, and challenges it might encounter.

We examined the commercial space-based positioning, navigation, and timing (PNT) market (an emerging market) and the commercial satellite communications (SATCOM) market (an established market) to evaluate the potential operational value that commercial offerings might add and the potential risks and challenges that the DAF might face when acquiring, integrating, and using these services for DoD operations. Our findings may also provide insights into potential opportunities, risks, and challenges that may arise in other emerging and established markets that have similar characteristics to the commercial space-based PNT and SATCOM markets. These findings may be of interest to USSF leadership, space acquisition authorities, and commercial space firms and trade groups.

The research reported here was commissioned by the Office of the Assistant Secretary of the Air Force for Space Acquisition and Integration (SAF/SQ) and conducted within the Force Modernization and Employment Program of RAND Project AIR FORCE as part of a fiscal year 2022 project, "Partnering with Commercial Space: Balancing Risks and Benefits."

RAND Project AIR FORCE

RAND Project AIR FORCE (PAF), a division of the RAND Corporation, is the DAF's federally funded research and development center for studies and analyses, supporting both the United States Air Force and the United States Space Force. PAF provides the DAF with independent analyses of policy alternatives affecting the development, employment, combat readiness, and support of current and future air, space, and cyber forces. Research is conducted in four programs: Strategy and Doctrine; Force Modernization and Employment; Resource Management; and Workforce, Development, and Health. The research reported here was prepared under contract FA7014-22-D-0001.

Additional information about PAF is available on our website:
www.rand.org/paf/

This report documents work originally shared with the DAF on October 6, 2022. The draft report, dated September 2022, was reviewed by formal peer reviewers and DAF subject-matter experts.

Acknowledgments

We would like to thank our sponsor, Brig Gen Steven Whitney, Military Deputy, SAF/SQ, for supporting this project and providing guidance. We are grateful to our project monitor, Col Jason Terry, Director of Architecture, Science and Technology, SAF/SQ, for his guidance, insight, and feedback throughout most of the project. His successor, Col Eric Felt, provided very helpful guidance in our formulation of recommendations. We also appreciate Chris Beauregard in SAF/SQ for providing his feedback and helping us connect with key subject-matter experts.

We thank many government subject-matter experts from the following organizations for providing insightful perspectives and data for this effort: Space Operations Command; Space Systems Command; the Space Warfighting Analysis Center; the Office of the DoD Chief Information Officer; the Space Rapid Capabilities Office; SpaceWERX; the National Reconnaissance Office; the Office of Space Commerce; SAF/SQ, Headquarters USSF, Office of Strategy, Plans, Programs, and Requirements; the Office of the USSF Chief Information Officer; Headquarters, U.S. Marine Corps Forces, Pacific; Headquarters, U.S. Indo-Pacific Command; the Combined Force Space Component Command; the Army Futures Command Assured PNT Cross Functional Team; and the U.S. Army Satellite Operations Brigade. Our work also benefited greatly from discussions with many commercial space industry members. We thank them for their contributions.

At the RAND Corporation, Moon Kim, Stephanie Pillion, and Natalia Henriquez Sanchez provided critical research support during the project. Air Force Fellow Lt Col Ryan Thulin and Army Fellow MAJ Eliot Proctor provided inputs based on their operational expertise, which greatly informed our research. We are grateful to Sherrill Lingel and Elizabeth Bodine-Baron, our program director and associate program director, respectively, for their helpful guidance and feedback throughout the project. We also thank our RAND colleagues Jim Bonomo, Mel Eisman, Michael Kennedy, and Richard Mason for their constructive comments during the research process. The content and recommendations of this report, however, are the responsibility of the authors.

Summary

Issue

The expanding commercial space industry offers services that could help the Department of Defense (DoD) meet its mission requirements more efficiently. These services range from commercial satellite communications (COMSATCOM) and remote sensing to collecting weather data; positioning, navigation, and timing (PNT); and space domain awareness. To better position itself to take advantage of these services, the Department of the Air Force (DAF) asked RAND Project AIR FORCE to help identify opportunities, risks, and challenges that the DAF might encounter when leveraging commercial space services in select mission and capability areas.

Approach

We focused our analysis on two space services: COMSATCOM and PNT. These were chosen because they represent the widest variation in market and firm maturity. We consulted literature on the use of commercial services to fulfill military needs, and we conducted more than 30 semi-structured interviews with subject-matter experts and stakeholders to develop an assessment framework for commercial services. This framework focuses on four broad categories of risk—operational, commercial market, commercial firm, and integration—to characterize and identify risks the DAF might encounter when leveraging commercial space capabilities and integrating them into the DoD enterprise.

Key Findings

- *The DAF seeks to fulfill four strategic goals when leveraging commercial space capabilities.* Although the DAF has not articulated a definitive set of strategic goals for leveraging commercial space capabilities, our review of the literature revealed four distinct ones. First, the DAF seeks increased resilience and mission capabilities. Second, it seeks to acquire them cost-effectively. Third, in the long term, the DAF seeks to foster a healthy space industrial base. Finally, the DAF seeks access to innovation spurred by commercial efforts. We found no strong order or hierarchy to these goals.
- *Commercial space services can provide additional capacity and resilience to existing space capabilities or provide new ones.* The commercial space-based PNT market could provide greater accuracy and signal strength than is currently available via GPS. The COMSATCOM market offers a variety of technologies that provide high throughput, jamming resistance, low latency, and global coverage. These services can be vulnerable to jamming, cyberattacks, and anti-satellite and ground network attacks—operational risks that should be considered when making decisions to use these services. However, leveraging services from multiple commercial service providers that use dissimilar or

heterogeneous technologies (and have dissimilar vulnerabilities) adds resiliency, which in turn mitigates many of these risks, as can adoption of new tactics, techniques, and procedures.

- *However, these services present new integration challenges.* Gaining operationally useful capabilities from commercial space services will require the DAF to remove barriers that limit commercial services from being fully integrated into military operations. Among other things, the DAF must articulate and implement hybrid architectures of commercial and military space capabilities that are developed on different timelines. More importantly, DoD's current ground and user segments for PNT and satellite communications (SATCOM) need to be modernized to enable broader access to commercial services. For example, PNT signals might need to be accommodated and integrated into user equipment that uses only GPS today. Other integration barriers, such as restrictions on the use of commercial services, also must be considered.

- *Market maturity will have the biggest influence on the ability of the DAF to meet its strategic goals.* COMSATCOM exemplifies a mature market; commercial demand is robust, as are resulting revenue streams, and sufficient financial stability allows providers and investors to explore new technologies. A foundation of a mature market is a customer base that includes not only DoD but also other government and commercial customers to spur innovation and vitality.

- *Investing in capabilities from an immature market puts the DAF's long-term goals at risk.* The DAF needs to be aware of the implications of investing in immature markets and measure its investment strategy against potential risks. For example, even though the long-term viability of the space-based commercial PNT market is limited, the DAF may judge that the immediate benefits of the capability to augment GPS are sufficiently valuable to pursue. But in making such investments, the DAF should be aware of and consider the potential impacts on long-term goals—such as industrial base stability—in a particular market sector.

Recommendations

- The DAF should be more willing to leverage commercial space services when market risks are low or can be spread among multiple stakeholders, the operational utility of the commercial service is high, and the integration risk can be managed.
- The DAF should adopt a wait-and-see approach using more-limited investment strategies, such as prototypes and pilot programs, when market conditions and the operational utility of the commercial service are uncertain.
- Regardless of its risk tolerance, the DAF should make the following refinements to the way it acquires commercial space services:
    - *Invest in greater market intelligence capabilities* to be able to stay abreast of developments in technical capabilities, as well as financial viability and market dynamics.
    - *Increase the sophistication of contracting capabilities* to be more adept at negotiating contracted services.
    - *Build flexible resourcing options* so that service contract negotiations can be conducted in a more timely fashion.

Contents

About This Report iii

Summary v

Figures and Tables ix

Chapter 1. Introduction 1
    The Department of the Air Force's Interest in Leveraging Commercial Space Services 2
    Project Objectives, Scope, and Methodology 2
    Organization of This Report 3

Chapter 2. The Department of the Air Force's Interest in Commercial Space Services 5
    The Continued Growth and Diversification of the Commercial Space Sector 5
    The Department of the Air Force's Implicit Prioritization of Four Distinct Goals for Using
        Commercial Space Services 8
    Sources of Risk That May Impede the Achievement of Those Goals 9
    An Analytic Framework 11

Chapter 3. Commercial Space-Based Positioning, Navigation, and Timing: A Limited
    Opportunity 13
    Potential Military Utility of the Space-Based Positioning, Navigation, and Timing Market 13
    Potential Difficulty in Mitigating the Operational Risks of Commercial Positioning, Navigation, and
        Timing 15
    The Cost and Complexity of Integrating Commercial Positioning, Navigation, and Timing at Scale ... 18
    The Uncertain Viability of Start-Up Firms in the Commercial Positioning, Navigation, and Timing
        Market 19
    Leveraging Nascent Commercial Positioning, Navigation, and Timing Services: A Limited
        Opportunity for the Department of the Air Force 23

Chapter 4. Commercial Satellite Communications: An Opportunity for Added Capacity and
    Resilience 24
    Stable Growth in the Satellite Communications Market 24
    Operational Benefits of Capacity, Diversity, and Innovations in the Satellite Communications
        Market 27
    Mitigation of Operational Risks Associated with Using Commercial Satellite Communications 29
    The Need to Modernize the Department of Defense Satellite Communications Ground Enterprise 33
    The Current, Fragmented Approach to Acquiring Commercial Satellite Communications 36
    Lack of Maturity in Integrated Satellite Communications Operations Management 38
    Summary 39

Chapter 5. Conclusions and Recommendations 41
    Commercial Space Services Can Provide Operationally Useful Capabilities 41
    Commercial Space Services Present New Integration Risks 42

Market Maturity Will Have the Biggest Influence on the Ability of the Department of the Air Force
    to Meet All Its Commercial Goals 42
Investing in Capabilities from an Immature Market Risks the Department of the Air Force's Long-
    Term Goals 43
Recommendations for the Department of the Air Force 44
Continued Growth of Commercial Space 45
Appendix A. Remote-Sensing Case Study 47
Appendix B. Complementing an Inertial Measurement Unit with Commercial Positioning,
    Navigation, and Timing..50
Appendix C. Risk Landscape for Proliferated Low Earth Orbit 52
Appendix D. Interview Protocols 54
Appendix E. Affiliations of Experts and Stakeholders Interviewed 56
Abbreviations 58
References 60

Figures and Tables

Figures

Figure 3.1. Positioning, Navigation, and Timing Performance Comparison: Commercial Service Versus GPS III with Military GPS User Equipment 17

Figure 3.2. Global Navigation Satellite System Augmentation Market Addressable by Commercial Positioning, Navigation, and Timing 22

Figure 4.1. Elements of U.S. Department of Defense and Commercial Satellite Communications Ground and Terminal Enterprises 34

Figure B.1. Potential Inertial Measurement Unit Performance Improvement 51

Tables

Table 2.1. Recent Changes in the Commercial Space Industry and Factors to Watch 6

Table 3.1. Potential Commercial Positioning, Navigation, and Timing Approaches to Help Mitigate Risk 16

Table 3.2. Projected Space-Related Start-Up Funding Available for Commercial Positioning, Navigation, and Timing 21

Table 4.1. Total Annual Revenues for Major Commercial Satellite Communications Companies 26

Table 4.2. Emerging Commercial Satellite Communications Technologies That the U.S. Department of Defense Can Leverage 29

Table 4.3. Emerging Commercial Satellite Communications Architectures That the U.S. Department of Defense Can Leverage 29

Table 4.4. Assessment of Commercial Satellite Communications Threat Mitigations: Technology-Based Mitigation Approaches 31

Table 4.5. Assessment of Commercial Satellite Communications Threat Mitigations: Architecture-Based Mitigation Approaches 32

Table E.1. Interviewees' Affiliations and Areas of Expertise 56

Chapter 1. Introduction

The expanding global commercial space industry offers new opportunities for the U.S. Department of Defense (DoD). Traditional commercial space markets, such as the satellite communications (SATCOM) and remote-sensing markets, are diversifying with new entrants that offer differentiated capabilities.[1] Technological advances in small satellites, lower launch costs, wider application of advanced manufacturing, and innovative satellite applications may enable even more growth.[2]

In addition, commercial markets are emerging to provide higher quality of service in areas that have traditionally been served by government agencies, such as weather; positioning, navigation, and timing (PNT); and space domain awareness.[3] Other markets are emerging to serve a growing number of satellite operators who seek to minimize their capital expenditures during start-up and operations.[4]

Developments in both types of markets are also being driven in part by nontraditional space companies. Nontraditional companies are those that have some combination of traits like the following: having limited or no previous history of working with DoD, being a start-up that is focused on bringing a single product or service to market, or being substantially supported by funding other than sales and revenue (e.g., venture capital, private equity). The commercial market is fast moving, and these traits are not exhaustive; however, they accurately describe some of the most common attributes of the kinds of nontraditional companies that are starting to work with DoD.

These developments may help DoD meet its mission requirements more efficiently while increasing resiliency, enable the sustainment of a robust domestic space industrial base, and give DoD access to a broader innovation pool.[5] The Department of the Air Force (DAF), particularly the U.S. Space Force (USSF), is increasing its partnership activities with nontraditional

---

[1] Examples of satellite applications include low-latency and high-bandwidth SATCOM, communications for underserved or unserved regions, and high–revisit rate satellite imagery.

[2] Regarding lower launch costs, we specifically refer to the lower launch service costs per payload lift mass. For more information on commercial space industry trends, see Matthew Weinzierl and Mehak Sarang, "The Commercial Space Age Is Here," *Harvard Business Review*, February 12, 2021.

[3] Weinzierl and Sarang, 2021.

[4] Examples of these new markets include launch services for small payloads, ground services for data transport and satellite telemetry, tracking and command, and on-orbit satellite servicing for mission life extension.

[5] Sandra Erwin, "Military Buyers Challenged to Stay Up on the Latest Commercial Space Innovations," *SpaceNews*, May 18, 2022c.

commercial space companies to take advantage of these opportunities in support of space operations.[6]

The Department of the Air Force's Interest in Leveraging Commercial Space Services

Although DoD has been using commercial space capabilities, such as SATCOM, for years, the recent rapid growth of the commercial space industry has given rise to questions about how reliable and useful these capabilities might be. Among other things, DAF leaders have expressed concerns that commercial space companies may dissolve because their long-term business plans may be highly uncertain and that these companies may, for political reasons, deny offerings in wartime that DoD relies on.[7] Furthermore, there are concerns about unknown instances of foreign involvement that could place DoD at risk (e.g., the sale of critical commercial technology that DoD depends on). Finally, U.S. government regulatory frameworks and other processes for vetting contractors that serve DoD may not sufficiently reflect the new environment—in terms of threats, technologies, and the nature of strategic competition—to adequately evaluate risks associated with nontraditional providers.[8] That said, additional regulatory and other bureaucratic layers could further increase the barrier to entry for commercial firms and limit DoD's access to innovation in the commercial sector.

Project Objectives, Scope, and Methodology

In fall 2021, the Assistant Secretary of the Air Force for Space Acquisition and Integration (SAF/SQ) asked the RAND Corporation to consider these issues to help the DAF **develop a balanced approach to leveraging commercial space companies in select mission and capability areas**.

This report summarizes our efforts, which consisted of three substantive research tasks:

1. Characterize the opportunity for partnering with commercial space companies.
2. Characterize key risks and challenges of partnering with commercial space companies.
3. Evaluate the DAF's processes and policies for leveraging opportunities in the space sector while addressing risks and challenges.

---

[6] Examples include USSF partnerships with space domain awareness companies, such as ExoAnalytic Solutions and LeoLabs.

[7] Julia Siegel, "Commercial Satellites Are on the Front Lines of War Today. Here's What This Means for the Future of Warfare," Atlantic Council, August 30, 2022.

[8] See, for example, Hannah Duke, "On-Orbit Servicing," Center for Strategic and International Studies, September 16, 2021; and Sandra Erwin, "Air Force Seeking Commercial Technologies for Cislunar Space Operations," *SpaceNews*, December 12, 2019.

Given time and resource constraints, we limited the scope of the project in two ways. First, at SAF/SQ's request, we focused on commercial space services in the space domain.[9] Second, we worked with SAF/SQ to identify two exemplar types of space operations—PNT and SATCOM—through which to explore benefits and challenges of leveraging commercial space services.[10] These missions were chosen because they represent the widest variation in market and firm maturity. Commercial PNT service offerings are in nascent stages of development; as of this writing, only a handful of firms are in the earliest stages of bringing this capability to market. Conversely, commercial SATCOM (COMSATCOM) services are provided by a relatively mature and stable commercial market that has been in existence for decades. This variation between the two markets is useful for identifying distinct and cross-cutting themes that could be generalized to other commercial space markets not analyzed in this research.[11] The insights about the PNT and SATCOM markets themselves may also be of some benefit to the DAF, but the goal of this research was to focus on broad and generalizable results about leveraging commercial space capabilities.

We used several methods to conduct this research. To gain an initial understanding of the policy problem, we consulted relevant policy and literature on the use of commercial services to fulfill military needs. We then conducted more than 30 semi-structured interviews with a variety of PNT and SATCOM subject-matter experts and stakeholders.[12] We thematically coded and analyzed these interviews to deepen our understanding of the benefits and challenges of leveraging commercial space services.[13] Where appropriate, we conducted independent market and technical analyses to verify and expand on interview themes. Interview protocols can be found in Appendix D, and a list of organizations interviewed can be found in Appendix E.

Organization of This Report

This remainder of this report consists of several chapters:

---

[9] As a result, we did not examine other opportunities for partnering with commercial space entities, such as hosting payloads on commercial satellites, transferring technology, or leveraging commercial-off-the-shelf satellite buses or components.

[10] DoD defines ten types of space operations in Joint Publication 3-14: space situational awareness; space control; PNT; intelligence, surveillance, and reconnaissance; SATCOM; environmental monitoring; missile warning; nuclear detonation detection; spacelift; and satellite operations (Joint Publication 3-14, *Space Operations*, Joint Chiefs of Staff, April 10, 2018, change 1, October 26, 2020).

[11] For more on case selection considerations, see Robert K. Yin, *Case Study Research: Design and Methods*, 5th ed., SAGE Publications, 2013.

[12] Stakeholders included DAF program managers involved in developing PNT and SATCOM capabilities, acquisition personnel focused on establishing partnerships with commercial providers, personnel involved in procuring and managing COMSATCOM contracts, and industry representatives.

[13] For a description of thematic coding, see Gery W. Ryan and H. Russell Bernard, "Techniques to Identify Themes," *Field Methods*, Vol. 15, No. 1, February 2003.

- Chapter Two explores the DAF's interest in leveraging commercial space services, particularly the DAF's strategic goals and the types of risks it may face.
- Chapters Three and Four discuss specific benefits and risks as they pertain to PNT and SATCOM, respectively.
- Chapter Five provides summative observations, findings, and recommendations to the DAF for a balanced approach to leveraging commercial space services.

Supporting information and methodological details are contained in the appendixes.

Chapter 2. The Department of the Air Force's Interest in Commercial Space Services

In this chapter, we explore the DAF's interest in leveraging commercial space services, particularly the DAF's strategic goals and the types of risks it may face.

The Continued Growth and Diversification of the Commercial Space Sector

RAND completed multiple studies in the past several years examining the commercial space industry and how DoD may leverage the capabilities that it offers. The most recent of these studies included data through December 2020.[14] Since that time, the capacity and capability in established sectors (e.g., SATCOM, space launch, remote sensing) have continued to grow.[15] Furthermore, growth in emerging markets and new entrants in established markets were largely driven by recent advances in small-satellite technologies, advanced manufacturing, artificial intelligence and machine learning, the adoption of proliferated constellations, and investments from venture capital firms.[16] Recent developments in these markets and key factors to watch for in the future are summarized in Table 2.1.

At the start of this project, we consulted the literature to assess the current status of these commercial space sectors and compare it with trends observed in previous studies. We confirmed that the global commercial space industry is largely continuing to grow and provide opportunities for DoD that are worthy of consideration. As far as the individual sectors are concerned, we did not see a significant change in the trends over the past two years, although there were a few minor exceptions.

Many commercial space sectors' new and recent entrants are taking advantage of small-satellite technologies, decreases in the costs of launch per payload lift mass, and the proliferated constellation model.[17] This is particularly prevalent in SATCOM and the emerging space-based low earth orbit (LEO) PNT markets, which we discuss in Chapters Three and Four, respectively.

---

[14] Yool Kim, George Nacouzi, Mary Lee, Brian Dolan, Krista Romita Grocholski, Emmi Yonekura, Moon Kim, Thomas Light, and Raza Khan, *Leveraging Commercial Space Capabilities to Enhance the Space Architecture of the U.S. Department of Defense*, RAND Corporation, 2022, Not available to the general public.

[15] See Emmi Yonekura, Brian Dolan, Moon Kim, Krista Romita Grocholski, Raza Khan, and Yool Kim, *Commercial Space Capabilities and Market Overview: The Relationship Between Commercial Space Developments and the U.S. Department of Defense*, RAND Corporation, RR-A578-2, 2022.

[16] Yonekura et al., 2022, p. v.

[17] BryceTech, *State of the Satellite Industry Report*, Satellite Industry Association, June 2021a; BryceTech, "Smallsats by the Numbers: 2021," presentation slides, August 13, 2021b, p. 11; Jeff Foust, "Launch Companies Optimistic About Future Demand," *SpaceNews*, September 9, 2021b.

Table 2.1. Recent Changes in the Commercial Space Industry and Factors to Watch

Sector	Changes in Recent Years	Factors to Watch
SATCOM	• Increased commercial capacity with increased market demand	• Added global broadband capacity from non-geostationary orbit satellite constellations

Space launch	• Increase in the number of launch-service providers across all launch classes	• Technology developments
		  – Reusability
		  – On-orbit reignition
		  – Increased lift capacity
		• Impact of National Security Space Launch Phase 2 contract award on market

Remote sensing	• Expansion in current and planned pLEO launches	• Size of commercial market and financial viability of start-ups
Environmental monitoring	• National Oceanic and Atmospheric Administration and DoD focus on Global Navigation Satellite System (GNSS) Radio Occultation (GNSS-RO) • Success in some GNSS-RO launch and operations, commercial and government	• New start-ups with developments in microwave, electro-optical and infrared, and space weather capabilities
Space domain awareness	• Increased demand with more entrants into the space domain	• Size of commercial market and financial viability of start-ups • Space proliferation driving demand and/or collaboration
Data transmit/receive networks (ground stations)	• New U.S. companies offering ground stations as a service for commercial and government customers	• Electronically steered antennas and multiphase array • Optical communications technology
Space logistics (on-orbit servicing)	• Launch of only one company • Developing niche capabilities from a few companies	• Realization of technological developments, enabling on-orbit refueling, assembly, and manufacturing

SOURCE: Adapted from Yonekura et al., 2022, p. vi, Table S.1.

Even commercial markets offering services that have traditionally been provided by government agencies at no cost are continuing to grow because of their ability to provide comparable or higher quality of service in such areas as weather, PNT, and space domain awareness.[18] The future of these markets was an open question in the 2020 RAND Project AIR

---

[18] Frederic Lardinois, "Weather Platform ClimaCell Is Now Tomorrow.io and Raises $77M," Tech Crunch, March 30, 2021; Emily S. Nightingale, Bhavya Lal, Brian C. Weeden, Alyssa J. Picard, and Anita R. Eisenstadt, *Evaluating Options for Civil Space Situational Awareness (SSA)*, Institute for Defense Analyses Science and Technology Policy Institute, August 2016; Jeffrey Pierre, "After a Year of Deadly Weather, Cities Look to Private Forecasters to Save

FORCE research,[19] so it is notable that the number of firms providing these services continues to grow.[20] Some of these and additional markets are emerging to support the large numbers of satellites and satellite operators as the proliferated constellation model takes hold of the industry.[21]

The main difference between our findings and those in another recent RAND report is that growth has not been as robust for the SATCOM and space data-transfer service market.[22] Many ground service providers are increasingly leveraging existing infrastructure when augmenting their current capabilities, which has resulted in a decline in the demand for new ground equipment.[23] Additionally, the development of the ground segment has been lagging that of the space segment, with more emphasis being put toward on-orbit systems than toward supporting infrastructure.[24] Overall, the SATCOM market is stable, but the revenues for the major providers have been flat for the past few years because more of the market share seems to be shifting from geosynchronous orbit (GEO) to proliferated LEO (pLEO).[25] Nevertheless, the SATCOM market is still projected to increase considerably over the next decade.[26]

The continued technological advances in and overall growth of the commercial space sector may provide opportunities for DoD to enhance its resilience, meet a subset of operational requirements more quickly and easily, and support the domestic space industrial base. However, this will require keeping up to date on developments in at least seven markets and dozens of individual firms within them, which is a task best completed continuously given the rate of change in some of the emerging markets and for new entrants. DoD should update its knowledge of the commercial space industry regularly.

Lives," NPR, December 16, 2021; Jason Rainbow, "Tomorrow.io to Grow Weather Constellation Through SPAC Deal," *SpaceNews*, December 7, 2021b; Debra Werner, "Startups Map Out Strategies to Augment or Backup GPS," *SpaceNews*, August 4, 2021.

[19] Kim et al., 2022.

[20] Lardinois, 2021; Pierre, 2021; Rainbow, 2021b.

[21] Ian Christensen, "Trends and Developments in Commercial Space Situational Awareness," presentation slides for seminar, Secure World Foundation, April 7, 2021; Jeff Foust, "Satellite Operators Need More Accurate SSA Data," *SpaceNews*, September 16, 2021c.

[22] Yonekura et al., 2022.

[23] For details on how infrastructure-sharing for the ground segment is enabled by the cloud, see Northern Sky Research, *Satellite Ground Segment: Moving to the Cloud*, January 2021; and Dan Swinhoe, "Data Centers with Dishes: How the Cloud Is Driving a Merger Between Data Centers and Ground Stations," DCD, December 22, 2021.

[24] Northern Sky Research, 2021.

[25] Northern Sky Research, "NSR's Satellite Capacity Report Sees Industry Moving Past COVID-19 Contraction to Drive $207B in Revenue Amidst Competition, Innovation and Risk-Taking," press release, July 27, 2022.

[26] Northern Sky Research, 2022.

The Department of the Air Force's Implicit Prioritization of Four Distinct Goals for Using Commercial Space Services

To consider the risks and benefits of leveraging commercial space services more fully, it is important to explicitly characterize the DAF's strategic goals for these services. Because the DAF has yet to release a definitive commercial space strategy that outlines these goals, we reviewed a variety of policymaker statements, strategic guidance (including national guidance), and other relevant sources to infer them.[27] Our review of these sources uncovered four consistent goals:

- **Access to innovation.** The DAF recognizes that the commercial space market is growing rapidly, creating new opportunities for innovative approaches and technologies that go beyond what the government-dominated legacy space market can conceive of. DAF leaders note that these technological advancements are being developed at a very high speed, which they believe can be used as a competitive advantage over rivals.[28]

- **A healthy industrial base.** DAF leaders have become increasingly aware of and uncomfortable with the prospect of a space industrial base that consists of only a few providers—and sometimes only a single provider. DAF leaders are keen to avoid relying on a key provider that is underperforming or might exit the market.[29] In either case, an anemic industrial base will make it difficult for the DAF to reliably maintain access to needed capabilities within reasonable cost, schedule, and performance bounds. The erosion of the industrial base can be mitigated by working with a larger number of nontraditional providers. Leveraging commercial space services also supports the long-term health of the industrial base more broadly by providing early funding for firms while they establish themselves in the commercial market.

- **Increased mission capability and resilience.** With rapidly growing threats in all domains, demand for DoD and USSF space capabilities that support a wide variety of user needs is on the rise. Furthermore, the USSF faces the challenge of ensuring that critical space capabilities are available in contested, degraded, and operationally limited environments. The DAF is looking to leverage the commercial space industry to augment DoD's space architecture and move toward a hybrid architecture that comprises commercial and DoD space systems in order to more reliably meet mission requirements. To that end, the DAF seeks to leverage a diverse set of existing and emerging commercial space capabilities to meet DoD's capability needs and enhance resilience to meet mission requirements.

---

[27] See, for example, John W. Raymond, *Chief of Space Operations' Planning Guidance*, U.S. Space Force, November 2020; Nathan Strout, "Space Force Will Set Up One Office for Commercial Services, Including SATCOM and Satellite Imagery," C4ISRNET, June 2, 2021; DoD, *Defense Space Strategy Summary*, June 2020; and White House, *National Space Policy of the United States of America*, December 9, 2020.

[28] In particular, in written testimony provided to Congress, DAF acquisition leaders state the goal that the DAF "will exceed the pace of demand and become increasingly more agile than our nation's adversaries" (Frank Calvelli and David D. Thompson, "Fiscal Year 2023 Priorities and Posture of the U.S. Space Force," statement presented to the Subcommittee on Strategic Forces, U.S. Senate, May 11, 2022, p. 9).

[29] Christian Davenport, "Disrupted by SpaceX, ULA Was in 'Serious Trouble.' Now It's on the Road Back," *Washington Post*, June 22, 2022.

- **Cost-effectiveness.** Related to the other goals is a desire to leverage the fruits of commercial research and development and economies of scale that became possible only in recent years. DAF leaders perceive that leveraging commercial services when appropriate can serve as a cost-effective substitute for building bespoke military capabilities.[30] This frees up government investment resources for other priorities.

Sources of Risk That May Impede the Achievement of Those Goals

The DAF's commercial space goals cannot be achieved without risk. We examined the literature on DoD's past experiences partnering with the commercial space industry, reviewed past research on benefits and risks of commercial space capabilities, and collected insights about potential risks from government and industry subject-matter experts.[31] We identified four broad categories of risk that were of concern to stakeholders: operational, commercial market, commercial firm, and integration.

Operational Risks

Some DAF PNT and SATCOM program personnel are concerned that the use of commercial space services might introduce risks to the successful accomplishment of DoD operations. The most-prominent operational risks we heard about from stakeholders are as follows (although others may exist):

- **Commercial services may not always be available.** Outsourcing a capability to a commercial provider raises the possibility of that provider denying services to DoD in times of conflict, possibly because of foreign government influence, a desire to avoid liability, or a desire to avoid associating itself or its other customers with conflicts. Firms may also wish to exit the market to pursue other business priorities.
- **Commercial services may be more vulnerable than government-provided services to adversary exploitation.** Using a commercial service may increase the risk of DoD information or data being compromised or of losing access to the service. The security posture of a commercial space service provider may be sufficient for commercial users, but the provider's systems or networks may have vulnerabilities that could lead to

---

[30] Sandra Erwin, "Space Force Looking to Ease Barriers to Entry for Commercial Companies," *SpaceNews*, April 4, 2022b.

[31] Appendix A contains a case study on remote sensing that helped guide the direction of our research. Prior RAND research includes Brien Alkire, Jonathan Fujiwara, Moon Kim, Yool Kim, George Nacouzi, Colby Steiner, and James Williams, *Leveraging Commercial Space Internet Services for Air Missions*, RAND Corporation, 2020, Not available to the general public; Michael Kennedy, Yool Kim, Brien Alkire, Benjamin M. Miller, Stephanie Young, Therese Marie Jones, and Matthew Sargent, *Analysis of Commercial Space Capabilities: Leveraging New Space to Increase Resilience of DoD Space—Business Viability Analysis*, RAND Corporation, 2019, Not available to the general public; Yool Kim, Ellen Pint, David Galvan, Meagan Smith, Therese Marie Jones, and William Shelton, *How Can DoD Better Leverage Commercial Space Capabilities? Understanding Business Processes and Practices in the Commercial Satellite Service Industry*, RAND Corporation, 2016, Not available to the general public; Kim et al., 2022; and George Nacouzi, Yool Kim, Jake McKeon, Alex Sedlack, Karishma R. Mehta, Mel Eisman, and Myron Hura, *Acquisition of Future Proliferated Low Earth Orbit Programs: Potential Challenges in Acquiring and Sustaining Mega Constellations*, RAND Corporation, 2020, Not available to the general public.

temporary or permanent loss of capability in the face of higher-level threats (kinetic and nonkinetic). Commercial space service providers might not implement adequate protection of data, systems, facilities, or networks, which could lead to loss, degradation, disruption, manipulation, or denial of sensitive information.

Interviews with various stakeholders revealed that some operational risks may be acceptable and may vary depending on the mission, the phase of a conflict, and how commercial services are being employed (e.g., as a primary, alternate, contingency, or emergency solution). The interviewees also discussed potential options for mitigating these risks. For instance, operators could employ new tactics, techniques, and procedures or use multiple commercial providers.

Commercial Market Risks

The literature and our interviews revealed that the commercial market must be viable without extensive DoD support in order for the DAF to rely on it. If commercial demand for a service is not sufficient to support sustained revenue for more than one or two firms, adverse outcomes may result. A weak commercial market could result in higher costs to DoD, vendor lock stemming from lack of competition, the consolidation of the industrial base, and reduction in the innovative potential that comes from meeting commercial demand.[32]

Commercial Firm Risks

We observed that the level of risk that DoD might incur will vary depending on the commercial space firm's ability to successfully launch and sustain its business while effectively competing in the market. There are numerous examples of commercial space firms that have gone through bankruptcy or exited the market. Relying on services to meet a critical DoD need from a firm that gets acquired by a foreign entity or exits the market could leave DoD with capability gaps. That said, if the market is competitive and robust, with multiple vendors supplying services to DoD, this risk could be mitigated.

In general, new entrants are likely to face considerable challenges on the path to becoming commercially viable. These challenges include capturing sufficient market share, obtaining financing for capital-intensive projects, and navigating a complex legal and regulatory landscape, all while overcoming technical challenges in system development. Incumbents in established markets must deal with similar challenges, but to a lesser degree given their experiences and status in the market.

Some stakeholders raised concerns about new entrants, who tend to be unfamiliar with DoD needs and processes. These stakeholders were concerned that these firms may be deterred by the complexities involved in working with DoD, limiting DoD's ability to access a broader pool of innovation. Some stakeholders were concerned about firms that might not adequately accommodate DoD-unique needs (e.g., capabilities that only DoD may need and that do not have

---

[32] See Appendix A for details on DoD's past experiences in procuring commercial remote-sensing services.

much commercial value or that support sensitive military applications).[33] Offerings from such firms may have limited operational value to DoD or introduce availability risk if such firms deprioritize or deny services to DoD for financial or political reasons. However, the literature suggests that if DoD becomes a dominant customer for a commercial firm, it could adversely affect the firm's ability to innovate and remain competitive in the commercial market.[34]

Integration Risks

DoD must integrate commercial services into DoD ground infrastructure, equipment, and operations to fully operationalize commercial services. The literature and our interviews revealed that integration challenges have created barriers that prevent commercial services from being able to fully support DoD missions. The DoD ground architecture is designed for military space systems, and it was developed and expanded over a very long timeline with technologies that are not compatible with rapidly evolving commercial technologies. In some situations, DoD desires integration of multiple niche capabilities from different providers or interoperability between commercial and military systems.

Implementing such integration will require additional investments. For instance, DoD may need to build or modify existing ground gateways, user equipment, or other infrastructure. Alternatively, it may need to pay commercial entities to build or modify the same. Such accommodations could make it more costly for DoD to leverage commercial capabilities. Modifying any existing or planned infrastructure and equipment could also be a source of risk for those programs that are involved in acquisition or sustainment of the affected military systems. Furthermore, integrating commercial services into DoD operations may require overcoming other nonmateriel challenges (e.g., doctrine, organization, training, and policy). These challenges could potentially lead to inefficiencies in how DoD buys commercial space services.

An Analytic Framework

We used these four risk categories—operational, commercial market, commercial firm, and integration—as an analytic framework with which to conduct our assessments of the risks that DoD might encounter when leveraging commercial PNT and COMSATCOM services. In conducting our analysis, we treated these four risk categories as largely distinct, but we

---

[33] One example of this is commercial weather data. DoD requires cloud characterization data to support planning for imagery overflights, search and rescue missions, and other military-specific needs. As of this writing, there is no commercial market for cloud characterization data. See Sandra Erwin, "Space Force Signals Demand for Commercial Weather Data, but Will the Industry Deliver?" *SpaceNews*, January 17, 2022a.

[34] See, for instance, Arthur Fishman and Rafael Rob, "Product Innovation by a Durable-Good Monopoly," *RAND Journal of Economics*, Vol. 31, No. 2, Summer 2000; and Michael Reksulak, William F. Shughart II, and Robert D. Tollison, "Innovation and the Opportunity Cost of Monopoly," *Managerial and Decision Economics*, Vol. 29, No. 8, December 2008.

recognize that they are interrelated in many ways. Additionally, because there are ways to mitigate some of the risks, we included these considerations in our assessment. In the next chapter, we discuss the integrated findings of our analysis for commercial space-based PNT.

Chapter 3. Commercial Space-Based Positioning, Navigation, and Timing: A Limited Opportunity

DoD's interest in commercial PNT stems from a desire to develop a set of robust and resilient PNT capabilities to meet mission needs. DoD currently relies on GPS as the primary PNT solution, but GPS may be denied or degraded in contested environments.[35] To that end, DoD is interested in multiple PNT sources to ensure that reliable PNT solutions are available in all operating conditions. Space-based commercial PNT services are one potential alternative that can be part of a diversified set of PNT solutions that complement GPS and increase resilience.

However, the DAF continues to modernize the GPS enterprise (space, control, and user segments) in such areas as increased security and jam resistance to improve GPS performance for military applications. In that context, we find that commercial PNT services that are coming online offer limited benefits to DoD. Additionally, the use of commercial PNT signals will require complex and costly modifications to current GPS receivers and host platforms, and the costs of these modifications may outweigh their limited operational benefit. Moreover, the market is not yet mature enough to provide any certainty about its financial viability without considerable DoD or U.S. government support. In this chapter, we discuss the potential benefits and challenges of leveraging commercial PNT for DoD.

Potential Military Utility of the Space-Based Positioning, Navigation, and Timing Market

Emerging commercial PNT companies are aiming to provide improved service over GPS, independent from GNSS (e.g., GPS, Galileo) that uses LEO constellations. Broadly speaking, these emerging companies are aiming to deliver stronger, encrypted signals for robustness to interference and spoofing; enhance PNT availability in restricted terrain (e.g., urban, indoor); and improve accuracy and other features (e.g., increase the speed of signal acquisition) for a variety of traditional civil applications (such as agriculture, civil air navigation, surveying, and telecommunications) and emerging civil applications (such as self-driving cars).

Three U.S. companies are either providing or seeking to provide independent PNT services: Satelles, Inc.; Xona Space Systems; and TrustPoint.[36] Satelles is the only firm that currently

---

[35] Congressional Budget Office, *The Global Positioning System for Military Users: Current Modernization Plans and Alternatives*, Pub. No. 4192, October 2011.

[36] A few non-U.S. companies are also potentially interested in providing GNSS-independent PNT, including the United Kingdom's OneWeb and China's Geely Holding Group. OneWeb has stated that it aims to offer PNT services, but this might not be feasible until it fields its second-generation constellation (Jeff Foust, "OneWeb

provides PNT from LEO. It uses the paging channel on the Iridium satellite constellation to provide global coverage (including coverage of the poles) and broadcast timing and positioning signals in L-band (1,616 to 1,626 MHz).[37] Satelles has been offering its Satellite Time and Location (STL) services to DoD and other customers since 2016. The signal strength of STL is between 300 and 2,400 times stronger than that of GPS.[38] This, along with the narrow spot beams that the satellites use to enhance authentication capabilities, makes STL signals quite difficult to spoof.[39] STL's positional accuracy is rather poor (30–50 m) for military applications,[40] and it takes a significant amount of time to calculate (on the order of minutes).[41]

However, STL also provides timing signals with an accuracy of about 30–50 nanoseconds (compared with GPS's accuracy of less than 30 nanoseconds), and an STL receiver can determine the time in under 2 seconds.[42] STL's timing signals have been proven effective in the lower floors of a 17-story building, where GPS was unavailable.[43] If DoD is willing to consider purchasing timing signals separately from position and navigation signals, Satelles is one potential option for DoD to increase its resilience and provide jam- and spoof-resistant timing signals in GPS-restricted areas.

Xona Space Systems was founded in 2019 and plans to launch its initial service in late 2024 or early 2025.[44] Its proposed PNT constellation comprises 300 cubesats, and it successfully

Continues to Study Offering Navigation Services," *SpaceNews*, April 8, 2021a). Since Geely successfully launched nine PNT satellites in late May 2022, it has had the largest number of PNT satellites in orbit after Satelles. Because Geely is part of an auto company, its main focus is to provide navigation services for autonomous vehicles (Zhang Yan and Ryan Woo, "China's Geely Launches First Nine Low-Orbit Satellites for Autonomous Cars," Reuters, June 2, 2022).

[37] Jeffrey A. Sherman, Ladan Arissian, Roger C. Brown, Matthew J. Deutch, Elizabeth A. Donley, Vladislav Gerginov, Judah Levine, Glenn K. Nelson, Andrew N. Novick, Bijunath R. Patla, Thomas E. Parker, Benjamin K. Stuhl, Douglas D. Sutton, Jian Yao, William C. Yates, Victor Zhang, and Michael A. Lombardi, *A Resilient Architecture for the Realization and Distribution of Coordinated Universal Time to Critical Infrastructure Systems in the United States*, National Institute of Standards and Technology, Technical Note 2187, November 2021.

[38] Sherman et al., 2021.

[39] Richard Mason, James Bonomo, Tim Conley, Ryan Consaul, David R. Frelinger, David A. Galvan, Dahlia Anne Goldfeld, Scott A. Grossman, Brian A. Jackson, Michael Kennedy, Vernon R. Koym, Jason Mastbaum, Thao Liz Nguyen, Jenny Oberholtzer, Ellen M. Pint, Parousia Rockstroh, Melissa Shostak, Karlyn D. Stanley, Anne Stickells, Michael J. D. Vermeer, and Stephen M. Worman, *Analyzing a More Resilient National Positioning, Navigation, and Timing Capability*, Homeland Security Operational Analysis Center operated by the RAND Corporation, RR-2970-DHS, 2021, p. 183.

[40] Mason et al., 2021, p. 90.

[41] The positional accuracy is for stationary devices, as Satelles uses Doppler ranging to determine locations (Mason et al., 2021, p. 212).

[42] GPS.gov, "GPS Accuracy," webpage, last modified March 3, 2022; Satelles, "Satelles and NIST Team Up for Precision Timing," March 30, 2022; Sherman et al., 2021; Werner, 2021.

[43] Sherman et al., 2021.

[44] Peter Gutierrez, "Fleshing Out the Leo PNT Landscape," *Inside GNSS*, March 14, 2022.

launched its first test satellite on May 25, 2022, and is scheduled to launch its second in 2023.[45] Xona's accuracy goals are based on the technical specifications of providing navigation signals for autonomous vehicles (e.g., sub-10-cm positional accuracy[46]), but the company is targeting a broad customer base that includes government entities and parties interested in critical infrastructure, drones, weather, and agriculture.[47]

TrustPoint was founded in 2020 and, like Xona, plans to field a constellation of LEO cubesats to provide independent PNT services. TrustPoint also hopes to provide the position and timing accuracy needed for navigation services for autonomous ground vehicles.[48] The company is early in its development and is still finalizing its business plans, but it is also targeting a broad customer base that includes both government and commercial customers, and it is focused on emerging markets, including automated vehicles, drones, and augmented reality.[49]

Potential Difficulty in Mitigating the Operational Risks of Commercial Positioning, Navigation, and Timing

To understand the net benefit of using commercial PNT, we assessed the implications of its use on operational risks, especially in the context of contested environments. Interviewees highlighted concerns about the potential vulnerability of commercial systems to various threats: jamming, cyberattacks, space-based anti-satellite (ASAT), ground-based ASAT, and ground network attacks. Table 3.1 shows our qualitative assessment of how technologies and architectures that commercial PNT providers are using (or planning to use) could mitigate these threats. Given that commercial PNT cannot be used as a primary PNT source in combat-related operations (per Chairman of the Joint Chiefs of Staff Instruction 6130.01G),[50] we relied on our own expertise to assess the relative contribution of commercial PNT to mitigating these threats.

---

[45] Jason Rainbow, "Xona Space Systems Fully Funds GPS-Alternative Demo Mission," *SpaceNews*, September 22, 2021a; "Xona Space Systems Secures Million$$$ in Investments from First Spark Ventures + Lockheed Martin Ventures," Satnews, August 3, 2022.

[46] Werner, 2021.

[47] Xona Space Systems, "Xona Pulsar," webpage, undated.

[48] Werner, 2021.

[49] Werner, 2021.

[50] Chairman of the Joint Chiefs of Staff Instruction 6130.01G, *2019 Chairman of the Joint Chiefs of Staff Master Positioning, Navigation, and Timing Plan*, Joint Chiefs of Staff, 2019.

Table 3.1. Potential Commercial Positioning, Navigation, and Timing Approaches to Help Mitigate Risk

PNT Operations-Related Threats	Jamming	Cyberattack	Space-Based ASAT (DE, EW, KKV)	Ground-Based ASAT (DE, DA)	Ground Network Attack
Technology-based mitigation approaches					
Smaller spot beams	Medium	Low	Low	Low	Low
Encryption	Low	High	Low	Low	Low
Architecture-based mitigation approaches					
Multiple providers	Low	High	High	High	High
pLEO	Low	Low	High	High	Low

NOTE: DA = direct ascent; DE = directed energy; EW = electronic warfare; KKV = kinetic kill vehicle. *Low* = minimal or no contribution to mitigation; *medium* = some contribution; *high* = significant contribution.

Smaller spot beams provide some additional protection against jamming because the received signal power is stronger; however, the small beams do not offer any advantage against the other threats presented in Table 3.1. Encryption offers protection against cyberattacks only. Augmenting GPS with PNT services from multiple providers and pLEO constellations could help mitigate risks due to multiple threats. However, the capabilities from these providers are unlikely to help mitigate jamming threats. These capabilities are all operating in the same frequency band as GPS (L-band), and the modernized GPS (GPS III, along with Military GPS User Equipment [MGUE] receivers) is likely to be more jam resistant. As a result, if GPS III is jammed, the commercial PNT is also likely to be jammed. Having multiple providers (with independent systems) increases resilience against several types of attacks (cyber, ASAT, and ground network), since an attacker needs to affect the systems from all the providers to successfully disrupt the signal. Because of the large number of individual platforms, pLEO likely complicates ASAT attacks; however, it may be less effective against jamming, cyberattacks, and ground network attacks, since the platforms have similar vulnerabilities related to these threats.

Figure 3.1 shows the performance of one commercial PNT service (Satelles) compared with that of the new GPS III with spot beam mode, along with the improved MGUE, operating in a notional jammed environment.[51] The performance of commercial PNT in a jammed environment is inferior to that of GPS III combined with MGUE. However, we add the caveat that our results are based on the information we had on Satelles, only because we did not have any specific performance data on Xona and TrustPoint systems. Nevertheless, we assess that commercial signals will not be available to the user if GPS III is jammed, unless commercial PNT is able to

---

[51] The estimated performance exhibited by Satelles is similar to or better than that of other commercial companies we assessed. For information on GPS III, see Congressional Budget Office, October 2011.

provide more jam resistance than that from GPS III spot beam plus MGUE or uses a different frequency band from that of GPS.[52]

Figure 3.1. Positioning, Navigation, and Timing Performance Comparison: Commercial Service Versus GPS III with Military GPS User Equipment

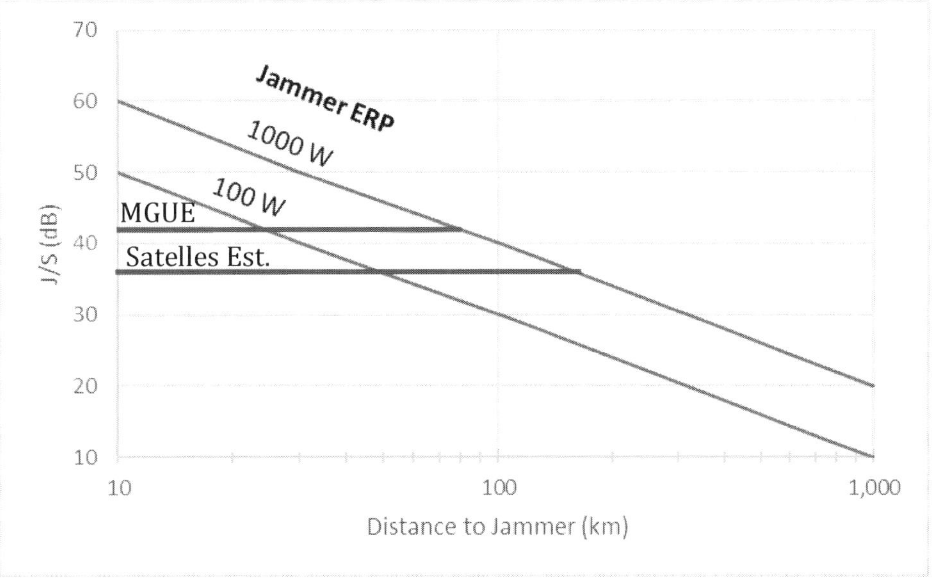

NOTE: dB = decibels; ERP = effective radiated power; Est. = estimate; J/S = Joules per second; W = watts. This figure shows the performance of a commercial PNT service (Satelles) compared with that of the GPS III with spot beam mode, along with the improved MGUE, operating in a notional jammed environment. The GPS III spot beam increases the terrestrial signal strength by about 10 decibel watts. This analysis assumes commercial receivers with acquisition J/s of 24 decibels (approximating today's military receivers) for Satelles.

Given this jamming vulnerability, relying on commercial PNT as the sole backup or complement to GPS carries risks. Depending on whether GPS is available, and depending on the criticality of PNT to the mission, the risk of using commercial PNT could vary from negligible to high. However, if commercial PNT is leveraged as one of many alternative PNT sources that complement GPS, the risk would be reduced. For example, the commercial signal can be used to calibrate and improve the inertial measurement unit (IMU) performance with minimal risk to the mission if properly blended into an integrated navigation system (INS) onboard the supported platform or weapon system (see Appendix B).[53] Additionally, we observed in our interviews with operational users that they were more comfortable with operational risks than other stakeholders were. They pointed to various tactics, techniques, and procedures to mitigate risk. They also pointed out that commanders bear the final authority to accept the operational risks mentioned here. Interviewees noted that, as long as the risks and their consequences are clearly

---

[52] Xona is considering C-band in the future, but, as we will discuss later in this chapter, that is likely to require a new antenna for the user equipment, potentially posing an integration risk.

[53] INSs are also known as *inertial navigation systems*.

articulated to them, they could identify appropriate uses for commercial PNT and ways to mitigate risks.

The Cost and Complexity of Integrating Commercial Positioning, Navigation, and Timing at Scale

Integrating commercial PNT services with DoD capabilities will likely be costly and complex. The literature and our stakeholder interviews highlighted four challenges in particular:

- implementing integration at scale
- aligning with ongoing GPS modernization efforts
- updating equipment with tight size, weight, and power constraints
- developing a commercial-military hybrid PNT architecture.

We examine each challenge in detail in the subsections that follow.

Implementing Integration at Scale

To some degree, the difficulty of integrating PNT at scale is a consequence of the success of the GPS program. GPS is a widespread program that is ubiquitous throughout DoD. For example, the Army currently has 500,000 GPS user equipment terminals and 250,000 precision weapons that are being considered for GPS upgrades. Replacing even a fraction of those receivers with ones that are compatible with commercial PNT signals would be a significant undertaking. This may be particularly complicated if a company like Xona uses C-band, requiring a separate antenna. Moreover, the ubiquity of GPS throughout DoD means that the integration process itself must run smoothly so as not to disrupt access to PNT services across the joint force.

Aligning with GPS Modernization Efforts

Current GPS modernization efforts are excellent examples of the scale that integrating PNT for DoD would entail. DoD is attempting to implement a new, stronger M-code GPS signal that is more resistant to jamming. The user equipment segment, known as *MGUE*, is expected to field an estimated 1.5 million pieces of user equipment between fiscal years 2017 and 2030.[54] However, the program has faced considerable obstacles, ranging from vendor hardware and software delays that have affected program schedules and disrupted test and evaluation efforts to delays with fielding plans and other integration-related issues. According to interviewees, DoD has fielded only several hundred MGUE devices. Furthermore, aligning commercial PNT service providers with the MGUE program would likely invite further programmatic risks.

---

[54] For additional details on modernizing or replacing receivers, see, for example, U.S. Government Accountability Office, *GPS Modernization: DOD Continuing to Develop New Jam-Resistant Capability, but Widespread Use Remains Years Away*, GAO-21-145, January 2021a.

Updating Equipment with Tight Size, Weight, and Power Constraints

Integrating commercial PNT signals into platforms with restrictive size, weight, and power constraints will be particularly difficult and will potentially require modifications to the host platform. Applications or algorithms that rely on PNT for precision timing (e.g., flight control or munitions avionics) would presumably need to be modified, further raising costs and programmatic risks. Conversely, potential changes in receiver size, weight, and power requirements may require modifications to the host platform itself.

Developing a Commercial-Military Hybrid Positioning, Navigation, and Timing Architecture

These integration problems are particularly challenging because they involve the integration of a *service* into DoD's PNT architecture. Different service providers may have different proprietary approaches, technologies, and interfaces with DoD equipment. The theoretical flexibility afforded by the acquisition of a service rather than a full, stand-alone capability is reduced by the friction of this transactional cost.

The ideal solution to this problem would involve a set of common standards or an open, software-defined architecture that can receive and integrate a variety of PNT signals alongside the existing GPS signal. For example, the U.S. Army is developing a software specification called the *PNT Operating System* (pntOS) *Modular Open Systems Approach* that developers can reference when developing PNT applications in order to reduce integration time for new PNT applications.[55] Such a hybrid architecture can be immensely useful, but it poses its own integration challenges. These might include defining the Modular Open Systems Approach, updating user equipment to accommodate it, and incentivizing commercial firms to adopt the standard.

The Uncertain Viability of Start-Up Firms in the Commercial Positioning, Navigation, and Timing Market

Although the potential benefits of commercial PNT services may be attractive to DoD, it is unclear whether a sustainable market will be established in the next five to ten years. Because there are so few firms offering services in the emerging space-based PNT market, the viability of the market is the same as the viability of the firms themselves. And the viability of U.S.-based companies that are providing or aim to provide commercial LEO-based PNT is likely to be endangered by regulatory and financial risks.

---

[55] For more details on the Modular Open Systems Approach and pntOS, see Open Innovation Lab, "Standards and Specifications," webpage, undated.

Emerging firms attempting to offer services via L- and C-bands have yet to secure the spectrum they need.[56] Representatives from these firms cited long delays and expenses as they waited for Federal Communications Commission (FCC) decisions on their filings. Moreover, firms providing PNT services outside the United States have access to 200 MHz in these bands while companies seeking to provide services in the United States can use only 120 MHz. This puts U.S.-based commercial PNT firms at a disadvantage. Without access to the spectrum in L- and C-bands, these firms will not be able to field their constellations and offer PNT services in the United States within the decade.

The most significant risk to the start-up firms, and, ultimately, to the sustainability of the market as a whole, is the uncertainty of immediate and long-term financial health. In the immediate term, commercial PNT firms are likely to struggle to attract start-up capital without significant DoD or other U.S. government support. Stakeholder interviews suggested that building and deploying a LEO constellation of approximately 300 PNT satellites will cost hundreds of millions of dollars, even before development and launch costs for the space segment are considered.[57]

However, according to Space Capital, a venture capital firm that invests in space technology companies, total venture capital funding for all space-related investments will be $20 billion in 2022.[58] If we assume that the current compound annual growth rate prevails for the next decade, we can assume that up to $34 billion in venture capital will be available in the next decade.

How much of that $34 billion might space-based commercial PNT firms attract? As of August 2022, one firm had raised $15 million in seed funding.[59] According to Space Capital, total seed funding in 2022 equaled approximately $750 million worldwide. Thus, that one firm attracted approximately 2 percent of all space-related seed funding. Even if we assume that space-based PNT becomes more attractive and attracts 5 percent of funding and that U.S.-based venture capital will continue to make up roughly half of all venture capital funding, one firm might expect to attract about $100 million in venture capital throughout all venture capital rounds associated with the premarket stage of a start-up. Although considerable, this amount is still short of the hundreds of millions of dollars of start-up capital needed to bring a commercial PNT service to market. Table 3.2 summarizes this analysis.

---

[56] Commercial PNT firm representative, interview with the authors, 2022.

[57] Jason Rainbow, "Lockheed Invests in Xona's GPS-Alternative Constellation," *SpaceNews*, August 3, 2022b. This figure is inferred from one firm's business plan to deploy a 60–200-satellite constellation consisting of low-cost payloads and commoditized buses at a cost of $6 million per satellite. See John Keller, "Northrop Grumman to Advance Position, Navigation, and Timing (PNT) for Secure Communications Satellites," Military & Aerospace Electronics, May 6, 2021.

[58] Space Capital, *Space Investment Quarterly: Q4 2022*, 2022.

[59] Rainbow, 2022b.

Table 3.2. Projected Space-Related Start-Up Funding Available for Commercial Positioning, Navigation, and Timing

Category	Estimate ($B)
2022 seed and Series A venture capital investment	2.6
2031 projected investment	4.5
2031 estimated U.S. share	2.2
2031 estimated U.S. PNT share	0.11

SOURCE: Features information from Space Capital, 2022.
NOTE: GNSS compound annual growth rate = 7 percent; U.S. venture capital share = 50 percent; PNT share = 5 percent.

This basic analysis is not meant to suggest that commercial funding is not possible in the commercial PNT market. Investor interest may increase if these firms develop paradigm-shifting technologies or approaches, or it may increase for some other reason. However, this analysis does highlight the distinct possibility that commercial PNT firms may not be able to attract enough venture capital funding to support the deployment of their constellations, necessitating DoD or other U.S. government funding (presuming that the government is committed to acquiring the capability). The DAF should pay careful attention to market conditions as it considers any commercial PNT investment opportunity.

The long-term prospects for stable revenue are also uncertain. We focused on estimating the total addressable market for commercial PNT by comparing our analysis of potential market segments with existing financial analyses of an existing service for which commercial PNT can be a substitute: GNSS augmentation.

According to the European Union Agency for the Space Programme (EUSPA), annual GNSS augmentation revenues will grow from $28 billion in 2022 to more than $50 billion by 2031. Thirty percent of the GNSS augmentation market comes from the automotive segment, which is driving the technological specifications of the start-up commercial PNT firms and is an important potential customer for these firms. An additional 3 percent of the market comes from segments that may benefit from the capabilities that a LEO PNT constellation may provide (e.g., increased signal strength, reduced need for extensive ground infrastructure): urban development, aviation, and emergency management.[60] If commercial PNT firms could capture even half of those market segments, they could expect to capture up to $9 billion in revenue in 2031 (see Figure 3.2).

---

[60] "Xona Space Systems Secures Million$$$ in Investments from First Spark Ventures + Lockheed Martin Ventures," 2022.

Figure 3.2. Global Navigation Satellite System Augmentation Market Addressable by Commercial Positioning, Navigation, and Timing

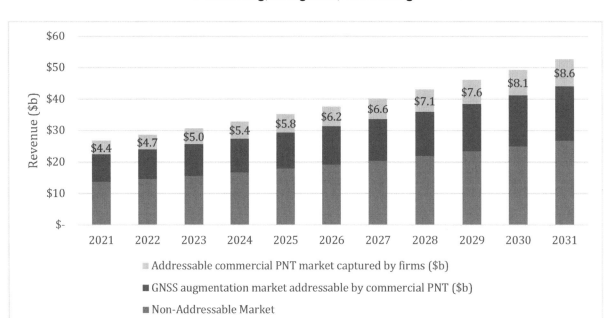

SOURCE: Features information from EUSPA, *EUSPA EO and GNSS Market Report*, No. 1, 2022.
NOTE: b = billions.

Although $9 billion in revenue seems sizable in absolute terms, much of this is predicated on EUSPA's forecast regarding the automotive market segment (30 percent), which might make use of commercial PNT for autonomous vehicle operation. However, autonomous vehicle technology is still in its infancy and can move in other technological directions that do not involve commercial PNT. Furthermore, it is unclear what the long-term fixed and variable costs of commercial PNT might be or how those services will be priced. These uncertainties cast some doubt on the adequacy of the revenue projections developed in this research.

Our assessment of these financial risks is based on the assumption that the commercial PNT firms will offer GNSS augmentation services as a way to generate revenue while they raise the funds they need to field their full constellations. It is possible that the firms will target additional markets, such as the GNSS hardware market, which is approximately twice the size of the GNSS augmentation market and is expected to grow at a similar rate.[61] Given the youth of the start-up firms and their developing business plans, it is difficult to assess whether even leveraging additional markets will provide the revenue they need to attain their planned capabilities within the next decade.

These risks are much less of a concern when considering the long-term financial stability of firms with existing constellations that might be repurposed for PNT services. This is the case

---

[61] EUSPA, 2022, p. 21.

with Satelles, which leverages an existing constellation and is currently generating revenue for its services, particularly its timing signals. Similarly, researchers have demonstrated the possibility of using SpaceX's Starlink constellation to provide PNT services.[62] The financial outlook for both cases is significantly more favorable owing to the fact that both firms have already launched constellations.

Leveraging Nascent Commercial Positioning, Navigation, and Timing Services: A Limited Opportunity for the Department of the Air Force

The potential for a complement or an alternative to GPS developed by commercial firms holds great interest for DoD as it continues to modernize to face great-power threats. However, the capabilities offered by firms in the market today provide limited military utility because of their vulnerability to jamming. If DoD decides that the capabilities offered today are worth the operational risks, our analysis shows that DoD will have to develop ways of integrating commercial PNT services into existing platforms and infrastructure. To do this with an acquisition system that is not optimized to stay aligned with fast-moving commercial technology will be especially problematic from a program management point of view.

Finally, if DoD decides that the capabilities are worth the risk and can integrate them into the force in a reasonable way, it likely will still have to provide substantial and ongoing financial support to ensure that commercial PNT capabilities continue to be available. Our analysis shows that, without a substantial increase in start-up funding and commercial demand, commercial PNT is not likely to be a viable market without considerable DoD or government support. Even if that support can be provided indefinitely, it will not meet DoD's strategic goals of broadening the industrial base and maintaining access to technological innovations from the commercial sector. Given this reality, the DAF may wish to influence how these capabilities evolve and build relationships to promote U.S. leadership in the commercial PNT sector. It can do this by supporting commercial PNT development efforts through existing defense innovation pathways (e.g., SpaceWERX, the Defense Innovation Unit) for low-risk missions or use cases. Doing so can provide initial start-up funding and help attract more commercial firms while providing a safe environment to test out capabilities and refine requirements. Moreover, such a relationship can keep lines of communication between industry and the DAF open to ensure that DoD's needs are understood as the market develops.

---

[62] Mark Harris, "SpaceX's Starlink Satellites Could Make US Army Navigation Hard to Jam," *MIT Technology Review*, September 28, 2020.

Chapter 4. Commercial Satellite Communications: An Opportunity for Added Capacity and Resilience

SATCOM is the oldest commercial space sector, dating back to the 1960s, when the first publicly traded COMSATCOM company was established following the Communication Satellite Act of 1962.[63] Since then, the industry has matured into a competitive commercial market with differentiated segments and strong motivation to innovate.

DoD has leveraged COMSATCOM to augment its military SATCOM (MILSATCOM) capacity for decades.[64] As DoD's on-orbit wideband SATCOM assets retire with no immediate follow-on programs of record defined, DoD's need for commercial capacity is anticipated to increase for the foreseeable future.[65] Additionally, as the threats to space capabilities increase, DoD is also looking for ways that COMSATCOM can help increase the resilience of the DoD SATCOM enterprise. At the same time, DoD is increasingly concerned about the vulnerabilities of COMSATCOM systems and services and the potential risks of using or relying on COMSATCOM, especially in times of conflict.

Fortunately, growing capacity, diversity, and innovations in the COMSATCOM industry over the years have the potential to meet many of these DoD needs. However, to take advantage of these benefits, DoD needs to mitigate potential operational risks associated with using COMSATCOM, modernize its SATCOM ground infrastructure and terminal enterprise, reform its acquisition practices for COMSATCOM, and establish centralized operations management of COMSATCOM. In this chapter, we discuss these benefits and challenges of leveraging COMSATCOM.

Stable Growth in the Satellite Communications Market

The overall SATCOM market, which offers consumer broadband, fixed-satellite services (FSS), and mobile-satellite services (MSS), appears stable. Global SATCOM market revenue recently decreased, from $24.4 billion in 2018 to $20.5 billion in 2020,[66] but, despite this

---

[63] Yonekura et al., 2022, p. 6.

[64] COMSATCOM augments wideband and narrowband MILSATCOM, and it does not provide protected SATCOM capability.

[65] There are currently no follow-on programs of record to Wideband Global SATCOM and the Mobile User Objective System.

[66] BryceTech, 2021a. Similar numbers were not included in the Satellite Industry Association's 2022 *State of the Satellite Industry Report* (BryceTech, *State of the Satellite Industry Report*, Satellite Industry Association, June 2022), so we are unable to report the 2021 values for these global revenues. Global revenues for FSS and MSS have

decline, the global market is expected to grow from $21.1 billion in 2021 to more than $30 billion in 2031.[67] DoD is a small part of this customer base, with less than 10 percent of global market share.[68]

Although there are more than 50 SATCOM providers worldwide, which offer a variety of SATCOM solutions for different applications (e.g., video distribution, defense, aerospace and maritime connectivity, cellular backhaul), the market share is concentrated in a smaller number of global operators. Major FSS operators are Eutelsat, Intelsat, SES, Telesat, and ViaSat, and major MSS operators are Inmarsat and Iridium. Eutelsat, Intelsat, SES, and Telesat used to make up 65 percent of the FSS market. However, market share for individual firms has shifted over the past decade, and their combined market share decreased to 37.6 percent. As of 2020, ViaSat has doubled its COMSATCOM market share since 2013 and is now the largest provider in that segment, up from fourth in 2018. Overall, revenues for these major COMSATCOM providers have been stable for the past two years, as shown in Table 4.1.

However, the market structure will continue to evolve because of recent mergers, such as ViaSat's acquisition of Inmarsat and the merger between Eutelsat and OneWeb.[69] Additionally, significant developments in architecture and technology in recent years have introduced new market dynamics. The two dominant trends are the addition of software-defined high-throughput satellites (HTSs) and the development of pLEO constellations. Through high frequency reuse and spot beam technology, HTS capabilities add substantial capacity to satellites (and reduce cost per bit) compared with traditional FSS GEO satellites. pLEO satellite constellations provide broadband services with reduced latency and increased resilience through proliferation and distribution. With these trends, SATCOM capacity supply is expected to grow substantially. The global HTS capacity supply is projected to grow over the next five years, exceeding 60 terabytes per second, and, as pLEO constellations are realized, the capacity supply will further increase dramatically.[70]

---

declined from $17.9 billion and $4.1 billion, respectively, in 2018 to $15.7 billion and $2 billion, respectively, in 2020. All figures are in then-year dollars. See BryceTech, 2021a.

[67] Northern Sky Research, 2022.

[68] This is consistent with DoD's consumption of COMSATCOM in terms of the global market and the total expenditures over the past seven years. See Commercial Satellite Communications Office, *Fiscal Year 21 Commercial Satellite Communications Expenditures and Usage Report*, Space Systems Command, U.S. Space Force, 2021, pp. 6, 10, Not available to the general public.

[69] OneWeb, "Eutelsat and OneWeb to Combine: A Leap Forward in Satellite Connectivity," press release, July 25, 2022b.

[70] Euroconsult, "High Throughput Satellites Poised to Become Leading Commercial Growers in Space Infrastructure, with Wholesale Capacity Revenues Projected to Top $100b by 2030," press release, March 23, 2022.

Table 4.1. Total Annual Revenues for Major Commercial Satellite Communications Companies

Company	Revenue (billions of dollars)				
	2013	2018	2019	2020	2021
Eutelsat	1.8	1.7	1.4	1.4	1.3
Intelsat	2.6	2.2	2	1.9	Not reported
SES	2.5	2.4	2	2	1.9
Telesat	0.9	0.7	0.7	0.6	0.6
ViaSat	1.1	1.6	2.1	2.3	2.3
Inmarsat	1.4	1.5	1.3	1.2	1.3
Iridium	0.4	0.5	0.6	0.6	0.6

SOURCES: Revenues for 2013 and 2018 are from Yonekura et al., 2022. Revenues for 2019–2021 are from Eutelsat Communications Group, *Consolidated Financial Statements as of 30 June 2020*, undated-a; Eutelsat Communications Group, *Consolidated Financial Statements as of 30 June 2021*, undated-b; Eutelsat Communications Group, *Consolidated Financial Statements as of 30 June 2022*, undated-c; Inmarsat Global Limited, *Annual Report and Financial Statements for the Year Ended 31 December 2020*, registered number 3675885, September 29, 2021; Inmarsat Global Limited, *Annual Report and Financial Statements for the Year Ended 31 December 2021*, registered number 3675885, September 30, 2022; Intelsat, *Annual Report Pursuant to Section 13 or 15(d) of the Securities Exchange Act of 1934 for the Fiscal Year Ended December 31, 2019*, U.S. Securities and Exchange Commission, commission file 001-35878, February 2020; Intelsat, *Annual Report Pursuant to Section 13 or 15(d) of the Securities Exchange Act of 1934 for the Fiscal Year Ended December 31, 2020*, U.S. Securities and Exchange Commission, commission file 001-35878, March 2021; Iridium Communications, *Annual Report 2019*, 2020; Iridium Communications, *Annual Report 2020*, 2021; Iridium Communications, *Annual Report 2021*, 2022; SES, *Consolidated Financial Statements as at and for the Year Ended 31 December 2019 and Independent Auditor's Report*, March 2, 2020; SES, *Consolidated Financial Statements as at and for the Year Ended 31 December 2020 and Independent Auditor's Report*, February 25, 2021; SES, *Consolidated Financial Statements as at and for the Year Ended 31 December 2021 and Independent Auditor's Report*, February 27, 2022; Telesat, *Annual Report Pursuant to Section 13 or 15(d) of the Securities Exchange Act of 1934 for the Fiscal Year Ended December 31, 2020*, U.S. Securities and Exchange Commission, commission file 333-159793-01, March 4, 2021; Telesat, *Annual Report Pursuant to Section 13 or 15(d) of the Securities Exchange Act of 1934 for the Fiscal Year Ended December 31, 2021*, U.S. Securities and Exchange Commission, commission file 001-39240 (Telesat Corporation) and commission file 333-255518-01 (Telesat Partnership LP), March 18, 2022; Viasat, *Annual Report Pursuant to Section 13 or 15(d) of the Securities Exchange Act of 1934 for the Fiscal Year Ended March 31, 2020*, U.S. Securities and Exchange Commission, commission file 000-21767, May 28, 2020; and Viasat, *Annual Report Pursuant to Section 13 or 15(d) of the Securities Exchange Act of 1934 for the Fiscal Year Ended March 31, 2022*, U.S. Securities and Exchange Commission, commission file 000-21767, May 27, 2022.
NOTE: Revenues are in then-year dollars.

GEO HTSs support a stable GEO market—adding capabilities rather than replacing them—and thus have a lower market risk, but the more nascent pLEO market is still shaping up and is therefore facing more uncertainty.[71] Although pLEO and other non-geostationary satellites were once seen as competition for GEO, these SATCOM segments appear to be working together and are sometimes offered by the same company. Eutelsat, Intelsat, Telesat, and ViaSat all have

---

[71] Euroconsult, 2022.

plans to provide services from a hybrid GEO/LEO network to provide the cost-efficiency of GEO and the low latency of LEO, depending on the applications being serviced.[72]

The overall SATCOM market is likely to remain robust and stable with diversified offerings. Its competitive nature is enabling innovations, as seen with the first wave of pLEO constellations coming online with Starlink and OneWeb. As of September 2022, Starlink and OneWeb alone have launched about 3,400 of more than 12,500 approved LEO SATCOM satellites.[73] Multiple firms have proposed tens of thousands more satellites for launch over the next ten years.[74] However, the long-term financial viability of pLEO is not yet proven, and pLEO operators need to overcome hurdles, such as securing access to the spectrum and financing their projects, in order to be successful. That said, the recent trend of pLEO providers working with or being acquired by traditional GEO providers to offer a complementary GEO/LEO capability mitigates certain risks for those pLEO providers. See Appendix C for more details on risks that the pLEO SATCOM segment faces.

Operational Benefits of Capacity, Diversity, and Innovations in the Satellite Communications Market

COMSATCOM capacity is expected to grow significantly and will far exceed that of MILSATCOM as providers continue to increase their capacity in different orbits at various frequencies (C-, Ku-, Ka-, and X-band). COMSATCOM satellites significantly outnumber MILSATCOM satellites in GEO and LEO, and the capacity of GEO HTSs greatly exceeds that of MILSATCOM.[75] With abundant capacity and proper user equipment, commercial providers could provide rapid surge capabilities.[76]

The number of COMSATCOM satellites in LEO is increasing at an astonishing rate. SpaceX is currently providing Ku-band global coverage using nearly 3,000 satellites and has plans to exceed 30,000 satellites over the next few years (and to use intersatellite links and other bands).[77]

---

[72] Jeff Foust, "GEO Operators Say They Can Compete Against LEO Systems on Cost," *SpaceNews*, March 22, 2022.

[73] OneWeb, "OneWeb Confirms Successful Launch of 34 Satellites, Delivering Ongoing Momentum at the Start of 2022," press release, February 10, 2022a; Mike Wall, "SpaceX Launches 46 Starlink Satellites, Lands Rocket at Sea," Space.com, last updated August 31, 2022.

[74] See, for example, Doug Messier, "Planned Comsat Constellations Now Exceed 94,000 Satellites," Parabolic Arc, November 8, 2021; Jonathan O'Callaghan, "The Risky Rush for Mega Constellations," *Scientific American*, October 31, 2019; and Michael Sheetz, "In Race to Provide Internet from Space, Companies Ask FCC for About 38,000 New Broadband Satellites," CNBC, November 5, 2021.

[75] For instance, DoD's Wideband Global SATCOM satellites have a capacity of less than 10 gigabytes per second (Gbps) each, while the Eutelsat KONNECT VHTS (very high–throughput satellite) is expected to provide a 500-Gbps capacity and ViaSat-3 is advertised to provide a 1,000-Gbps capacity per satellite.

[76] "A New Class of Protected Satellite Communications," blog post, Intelsat, May 13, 2015.

[77] Jason Rainbow, "Space X Requests Spectrum to Upgrade Starlink Mobile Services," *SpaceNews*, July 27, 2022a.

Other companies, such as Amazon, are planning to deploy their own constellations of communication satellites in LEO.[78] As these satellites proliferate, global coverage multiplies, providing worldwide users with additional SATCOM coverage. DoD can use these developing capabilities to further increase the available SATCOM coverage it can leverage.

The collection of COMSATCOM providers and satellites offers opportunities to access services in different orbits. If there is appropriate preplanning, users would be able to leverage additional constellations and satellites when they are needed, lowering risk from jamming attacks and improving overall resilience. New and future pLEO constellations provide a large number (we estimate greater than six satellites) of access opportunities at a given moment to users, while the multiple GEO COMSATCOM providers enable users to connect with different GEO satellites. These multiple access opportunities can improve the ability of a user to access SATCOM services even when an opponent is jamming some of the satellites.[79] Additionally, providers are planning to offer seamless multi-orbit services, which should further enhance connectivity to users.[80]

In addition to the capabilities just discussed, SATCOM providers are pursuing many other technologies and architectural approaches to provide improved and differentiated offerings, which may offer additional operational benefits to DoD. These offerings are summarized in Tables 4.2 and 4.3. Many of these capabilities also serve as mitigation measures to potential operational risks that DoD users may be concerned about. We discuss our assessment of these technologies in the context of mitigation measures in the next section.

---

[78] Jamie Morin and Robert S. Wilson, *Leveraging Commercial Space for National Security*, Center for Space Policy and Strategy, Aerospace Corporation, November 2020.

[79] An opponent would need to simultaneously track and jam several LEO satellites, as well as satellites in GEO and medium earth orbit (MEO), to deny a user SATCOM services.

[80] ThinKom PR, "SATCOM for U.S. DoD to Be Delivered by ThinKom, Inmarsat Government," press release, October 4, 2022.

Table 4.2. Emerging Commercial Satellite Communications Technologies That the U.S. Department of Defense Can Leverage

Technology Type	Description
Rapid ad-hoc networking	Temporary networks that enable communication between terminals without routers or other intermediaries
Transponder lease with protected tactical waveform (PTW)	Leased commercial transponders hardened with PTW
Small spot beams	Higher power signal concentrated in a specific region
Steerable beams	The ability of a satellite to control where it is providing coverage within a specific region
Beam forming and nulling	Signal filtering outside a desired spatial region that increases jamming resistance
Intersatellite links	Communication between satellites that increases resilience
Software-defined radios	Adaptive and reconfigurable systems enabled by software
High- and very high–throughput satellites	Satellites that can support significantly higher throughput capacity without requiring additional spectrum access
Frequency diversity	Greater diversity of frequencies utilized
Adaptive link power and bandwidth control	Bandwidth optimization that improves resource use efficiency

Table 4.3. Emerging Commercial Satellite Communications Architectures That the U.S. Department of Defense Can Leverage

Architecture Type	Description
pLEO	Proliferation and distribution that lowers latency and increases resiliency
Direct control of COMSATCOM resources	Direct control of purchased COMSATCOM resources that enables the ability to reallocate them where needed
Integrated multi-orbit service	Greater resiliency from a greater number of orbits
Multiple providers	A third-party integrator or partnership to offer an integrated SATCOM solution from multiple providers

SOURCE: Features information from Eutelsat, "Combining Satellite and OTT Delivery," brochure, undated.

Mitigation of Operational Risks Associated with Using Commercial Satellite Communications

Although COMSATCOM has the potential to provide many benefits, DoD would need to consider and mitigate potential operational risks when using these services, especially in the context of contested environments. Military users of COMSATCOM and subject-matter experts usually identify (1) availability, (2) cybersecurity, and (3) physical security as the principal operational risks related to the use of COMSATCOM.[81]

Many of these risks are being addressed by both the commercial provider and the USSF. The emerging technologies and architectural approaches discussed in the previous sections serve as potential mitigation measures. We qualitatively assessed these technological and architectural mitigations (available now or in the near future) that can be implemented within a SATCOM enterprise; the results of this assessment are depicted in Tables 4.4 and 4.5. The three operational risks mentioned above are affected by five principal threat vectors, which we list as the column headings in the tables: jamming, cyberattack, space-based ASAT, ground-based ASAT, and ground network attack. The ratings in the tables range from *high* to *low*, and they represent the capability of the approach or attribute to address the identified threats. The ratings are subjective by necessity and were developed through expert judgment relative to a traditional GEO-based COMSATCOM offering.

Anti-jam capabilities are improved with the use of flexible beams; adaptive link power and bandwidth control; integrated multi-orbit services; and tactics, techniques, and procedures to circumvent some of the jamming threats. Implementing PTW with COMSATCOM also provides jam-resistant communications to the user.[82] PTW achieves this technically by using frequency hopping, frequency diversity, National Security Agency cryptography, time-division multiple access, and time synchronization.[83] One of the main benefits of PTW is that it is developed by DoD to provide users with "a non-proprietary, flexible waveform capable of low probability of detection/intercept (LPI/LPD) and significant anti-jam performance with Suite B cryptographic support over a variety of satellite types and frequencies."[84] From DoD's perspective, this approach avoids vendor lock and minimizes requirement changes to COMSATCOM systems while simultaneously addressing concerns related to integrity and availability in the commercial market.[85] DoD should consider influencing some of the commercial offerings, as it did with

---

[81] *Availability* is the ability to obtain the service when needed; *cybersecurity* involves the integrity of the information; and *physical security* involves the security of the network, including the satellite and ground elements.

[82] Amanda Miller, "Jam-Resistant Waveform for Safer Battlefield Communications to Get First In-Space Test," *Air Force Magazine*, March 23, 2022.

[83] PTW further enables efficient resource allocation of SATCOM bandwidth by dynamically assigning the communication mode and assigning radio frequency resources on the basis of traffic demand without the need for a human in the loop. And higher data rates are achieved by maintaining smaller link margins. See Military Satellite Communication Analysis, Systems Integration and Test, and Engineering Services, "Protected Tactical Enterprise Service (PTES) - Protected Tactical Waveform (PTW) Interface Control Document (ICD)," Air Force Space and Missile Systems Center, MILSATCOM Systems Directorate (SMC/MC), March 9, 2018, Not available to the general public.

[84] Brian Peterson, "Protected Tactical Waveform (PTW) Engineering Overview," Headquarters Air Force Space Command/A5M, March 16, 2016, Not available to the general public.

[85] Lisa A. Baghal, *SATCOM Supply Versus Demand and the Impact on Remotely Piloted Aircraft ISR*, Air Command and Staff College, Air University, March 2016.

PTW by supporting standard intersatellite link interfaces, which should further increase the value of COMSATCOM for other space-based missions.

Table 4.4. Assessment of Commercial Satellite Communications Threat Mitigations: Technology-Based Mitigation Approaches

COMSATCOM Operations-Related Threats	Jamming	Cyberattack	Space-Based ASAT (DE, EW, KKV)	Ground-Based ASAT (DE, DA)	Ground Network Attack
Rapid ad-hoc networking	Medium	Medium	Medium	Medium	Medium
Transponder lease with PTW	High	Low	Low	Low	Low
Small spot beams	High	Low	Low	Low	Low
Steerable beams	Medium	Low	Low	Low	Low
Beam forming and nulling	High	Low	Low	Low	Low
Intersatellite links	Medium	Low	Medium	Medium	Low
Software-defined radios	High	Low	Low	Low	Low
HTS/VHTS	Medium	Medium	Medium	Medium	Medium
Frequency diversity	High	Low	Low	Low	Low
Adaptive link power and bandwidth control	Medium	Low	Low	Low	Low

NOTE: VHTS = very high–throughput satellite. *Low* = minimal or no contribution to mitigation; *medium* = some contribution; *high* = significant contribution. The ratings were developed through expert judgment relative to a traditional GEO-based COMSATCOM offering.

Table 4.5. Assessment of Commercial Satellite Communications Threat Mitigations: Architecture-Based Mitigation Approaches

COMSATCOM Operations-Related Threats	Jamming	Cyberattacks	Space-Based ASAT (DE, EW, KKV)	Ground-Based ASAT (DE, DA)	Ground Network Attack
Multiple providers	High	High	High	High	High
Proliferated LEO	High	Low	High	Low	Low
Integrated multi-orbit service	High	Low	High	Low	Low
Direct control of COMSATCOM resources	Medium	Medium	Medium	Medium	Medium

NOTE: *Low* = minimal or no contribution to mitigation; *medium* = some contribution; *high* = significant contribution. The ratings were developed through expert judgment relative to a traditional GEO-based COMSATCOM offering.

Some of these commercial providers, such as Starlink, have some inherent resilience based solely on the large number of satellites orbiting in LEO and narrow beam widths. pLEO satellites afford users the advantage of being able to access several satellites at any given time, which complicates an opponent's plan to deny users SATCOM services.

To further mitigate cybersecurity risks, the USSF deployed a cybersecurity initiative, the Infrastructure Asset Pre-Approval program, to improve the cybersecurity posture of COMSATCOM providers supporting DoD.[86] The program involves two steps: First, the COMSATCOM provider receives a cybersecurity assessment by a third-party assessor certified by the USSF. The second step is a subsequent review of the assessment results by the USSF for approval. If approved, the provider and the assessed asset will be placed on an approved list for potential future contracts. To address cybersecurity risks, some providers are hardening their end-to-end networks with encryption and ground physical security.[87]

Tables 4.4 and 4.5 show that although one individual technological or architectural approach could mitigate operational risks against one threat or a subset of threats, it cannot mitigate all. However, leveraging multiple providers whose collective capability comprises heterogeneous technologies and architectures results in the best chance of mitigating operational risks against a variety of threats, as shown in Table 4.5.[88] Operational risks can be further mitigated by selecting

---

[86] Space Systems Command, "SSC CSCO Reaches Critical Milestone for IA-Pre, Roll-Out Begins Today," press release, May 26, 2022.

[87] "A New Class of Protected Satellite Communications," 2015.

[88] With respect to cybersecurity, we assess that leveraging multiple providers would contribute significantly to increasing resilience relative to leveraging a single provider, because it would be more difficult for an attacker to

providers who have implemented improved security measures and putting in place appropriate contractual vehicles to ensure that the service is available when it is needed.

However, DoD needs to also consider the criticality of the mission and the impact of COMSATCOM failure when deciding whether to leverage COMSATCOM services. Many missions (such as morale, welfare, and recreation and unmanned aerial vehicle communications) can tolerate some risk, while other missions (such as subsets of nuclear command, control, and communications) have lower risk tolerances, which might disqualify the use of COMSATCOM.

The Need to Modernize the Department of Defense Satellite Communications Ground Enterprise

As discussed earlier, risks from multiple threats could be mitigated by enabling users to access multiple providers' diverse technologies and architectures such that users can move to a different beam, satellite, or network as needed, making it difficult for an adversary to deny connectivity. However, DoD's current SATCOM ground infrastructure and user terminals are the biggest barriers to integrating COMSATCOM services derived from heterogeneous technologies, multiple orbits, multiple frequency bands, and multiple networks into DoD's SATCOM enterprise.

Figure 4.1 illustrates various elements of the SATCOM ground and terminal enterprises that need to be compatible with commercial networks to integrate COMSATCOM services into the DoD SATCOM enterprise. Significant investments are needed for an end-to-end integration between commercial and MILSATCOM architectures; integrations are needed at the terminal and platform, gateway, network, cloud site, and enterprise management levels for operational flexibility and resilience. Furthermore, different services and organizations own different elements of the ground architectures and terminals, which makes coordination and synchronization of investment decisions among these services and organizations extremely difficult.

---

deny a user's access to all of the providers' networks than for the attacker to deny the user's access to only a single provider's network. However, we assess that an architecture with multiple providers might not necessarily reduce the potential risk of losing confidentiality or integrity of user data, depending on the architecture and network management configuration. For instance, an attacker might need to exploit only one provider's network to compromise user data.

Figure 4.1. Elements of U.S. Department of Defense and Commercial Satellite Communications Ground and Terminal Enterprises

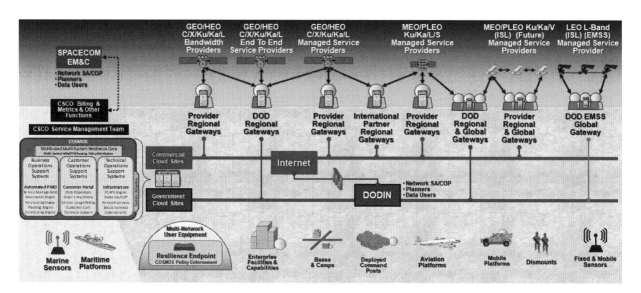

SOURCE: Reprinted from Jared Reece, "Infrastructure Asset Pre-Approval (IA-Pre) Kickoff," briefing slides for presentation to Commercial Space Cryptographic Cybersecurity Working Group, Space Systems Command, May 26, 2022.
NOTE: AAA = authentication, authorization, and accounting; COP = common operating picture; COSMOS = Comprehensive Open-Architecture Space Mission Operations System; CSCO = Commercial Satellite Communications Office; DODIN = DoD information networks; EM&C = enterprise management and control; EMSS = enhanced MSS; FCAPS = fault, configuration, accounting, performance, security; HEO = high earth orbit; ISL = intersatellite link; PMO = program management office; PPPP = public, private, people, partnership; SA = situational awareness; SPACECOM = U.S. Space Command.

Terminal modifications or replacements may be required to accommodate access to services from multiple COMSATCOM providers and to MILSATCOM systems. For example, some DoD platforms may be limited in the bands they can operate in, which limits their compatibility with commercial services outside those bands unless terminal modifications are made.[89] Some COMSATCOM providers (e.g., ViaSat) employ proprietary waveforms, which further limit terminal flexibility. DoD users can access a commercial network via DoD-unique theater terminals or via commercial terminals, but both approaches require DoD to procure terminal equipment that is compatible with commercial frequency bands and waveforms or modify existing equipment to access both military and commercial networks.[90]

Additionally, using commercial services from multiple orbits could require different antennas with separate terminals to access the various radio frequencies used by different providers. While

---

[89] For example, airborne intelligence, surveillance, and reconnaissance platforms, such as Global Hawk, were originally equipped with Ku-band terminals and could not use the Wideband Global SATCOM constellation, which operates in military Ka- and X-bands.

[90] See, for example, reporting on this from 2000 in Tim Bonds, Michael G. Mattock, Thomas Hamilton, Carl Rhodes, Michael Scheiern, Philip M. Feldman, David R. Frelinger, and Robert Uy, *Employing Commercial Satellite Communications: Wideband Investment Options for the Department of Defense*, RAND Corporation, MR-1192-AF, 2000.

users would like to integrate COMSATCOM services from LEO, MEO, and GEO, this may require multiple antennas because each orbit requires a different antenna focus to receive the signals from the satellite. Even within LEO, different antennas are required to catch the signals from both the rising and setting satellites.[91] Integrating a COMSAT terminal onto a DoD host platform could also be complex, depending on the platform. Ground-portable terminals are stand-alone and relatively straightforward to integrate, while integrating a new terminal and antenna onto an aircraft requires careful consideration of how to avoid interfering with existing systems on the platform, which has limited available size, weight, and power.[92]

Services are pursuing developmental efforts to field multiband terminals and terminals that can be used with both COMSATCOM and MILSATCOM systems.[93] However, modernizing the existing SATCOM terminals (and any associated integration onto host platforms) will require a significant investment and could be a lengthy process, as DoD has more than 17,000 terminals that represent about 135 different designs.[94]

In addition to the terminals, many other interface points between the commercial and military ground architectures need to be integrated to operationalize a hybrid commercial and military architecture. For example, COMSATCOM provides access to DoD information networks, such as the Defense Information System Network, for many DoD users. This requires commercial systems to be tied to the DoD ground entry points or gateways. Enhancements in DoD gateways are needed to increase the capacity to accommodate traffic to and from commercial systems.[95]

Tying all the different networks together will also require developing complex enterprise management and control capabilities to provide situational awareness and to facilitate creating a common operating picture, and it will require automated resource allocations to efficiently allocate satellite bandwidth and to switch users to different systems or networks as needed.[96]

---

[91] See, for example, Theresa Hitchens, "Antennas: The Hard Physics Challenge for Space Force 'Hybrid' SATCOM Plan," *Breaking Defense*, April 15, 2022.

[92] U.S. Government Accountability Office, *Satellite Communications: DOD Should Explore Options to Meet User Needs for Narrowband Capabilities*, GAO-21-105283, September 2021b.

[93] For example, the U.S. Army has a program called *Product Manager Satellite Communications* that manages multifrequency-band satellite terminals (C, Ku, Ka, X) to allow users to leverage both commercial and military communication satellites (U.S. Army, "Satellite Communications (SATCOM)," June 2, 2016). In addition, the GX30 airborne Ka-band terminal was recently approved for use over Inmarsat's Global Xpress network (which is designed to be compatible with Wideband Global SATCOM) for both military and commercial users of Ka-band. See Orbit, "Orbit's GX30 Terminal Receives Type Approval to Operate on Inmarsat Global Xpress Commercial and Military Ka-Band Networks," press release, August 16, 2022.

[94] U.S. Government Accountability Office, *Satellite Communications: DOD Should Develop a Plan for Implementing Its Recommendations on a Future Wideband Architecture*, GAO-20-80, December 2019.

[95] Note that DoD could outsource the gateway service from commercial entities and does not need to own and operate its gateways.

[96] According to DoD Instruction 8420.02, *DoD Satellite Communications*, U.S. Department of Defense, November 25, 2020, p. 28, *DoD SATCOM enterprise management and control* refers to

Currently, switching a user from one system or network to another would entail slow manual or paper-based processes.[97] The USSF is developing the Enterprise Satellite Communications Management and Control (ESC-MC) infrastructure to provide such capabilities. ESC-MC will require the development of many interfaces and standards to enable information-sharing at the network level between the centralized enterprise management and control system and SATCOM elements (military or commercial).

Coordinating and synchronizing all the integration activities pointed out here adds another layer of complexity because different organizations are responsible for various pieces. For example, the Defense Information Systems Agency is responsible for the gateways, the USSF is developing ESC-MC, and each service is developing and acquiring its own terminals.

The Current, Fragmented Approach to Acquiring Commercial Satellite Communications

The current approach to acquiring COMSATCOM is fragmented and does not lend itself to supporting the operational flexibility intended in an integrated military and commercial SATCOM enterprise with multiple providers.

DoD procurement of COMSATCOM services is currently centralized under the USSF via CSCO, per U.S. Space Command Instruction (SPI) 3250.01A, unless a waiver is issued.[98] COMSATCOM requirements from the requesting DoD organization (which also provides the funding) arrive at CSCO, and CSCO examines each request and procures COMSATCOM

---

the DoD SATCOM enterprise top-level management and control system that uses a service-oriented architecture to provide access to DoD SATCOM enterprise [information technology] services (to one or more SATCOM element networks as well as to various pre-defined user/manager accounts) or assists networks with scheduling and requesting resources from another element service provider. [It] supports the business and operational functions of the seven core DoD SATCOM enterprise management and control capabilities (integrated data storage management, provide SA [situational awareness] information, enterprise network management to element management communications, identify and access management services, DoD SATCOM enterprise modeling and analytics, network access control, and multi-vendor network element management service).

[97] Theresa Hitchens, "Space Force to Focus SATCOM Management on JADC2 Needs: Exclusive," *Breaking Defense*, December 18, 2020.

[98] SPI 3250.01A, *Satellite Communications*, U.S. Space Command, May 18, 2020. By law, DoD components procuring COMSATCOM services must go through CSCO unless there is a procurement waiver, which CSCO can facilitate, and the DoD Chief Information Officer decides to allow the waiver. These waivers are generally uncommon, but interviewees noted that some customers bypass CSCO. For instance, because the requirement to go through CSCO applies specifically to operational needs, customers can circumvent CSCO by entering into research and development contracts with COMSATCOM providers. Interviewees indicated that customers may circumvent CSCO to potentially avoid fees or because they perceive that CSCO is not being responsive to their operational needs. We did not have any information on how often or how many contracts are executed by organizations other than CSCO. As a result, there may be gaps in CSCO's awareness of commercial SATCOM usage by DoD.

solutions to meet those requirements on an individual basis (i.e., a one-on-one contract between the DoD customer and the COMSATCOM provider).

Each contract is negotiated to meet the requesting organization's COMSATCOM requirements based on their mission need, which can vary in terms of the complexity and the duration of the COMSATCOM solution (e.g., for training, for meeting a short-term bandwidth gap, or for an end-to-end managed service). The contracting vehicle and contractual terms vary depending on the users' COMSATCOM requirements, and the resulting COMSATCOM solution may entail user-unique network configurations and terminals. These customized solutions limit users from shifting SATCOM resources as their mission needs change, unless the contract is modified or a new contract is negotiated.[99] Interviewees indicated that these contracts are primarily for supporting operations in a benign environment, as COMSATCOM services tend to be purchased for short-term leases (partly because of appropriation limitations for SATCOM leases).

As a result, DoD is essentially procuring COMSATCOM via a large number of disparate, short-term contracts,[100] many of which may provide overlapping capabilities, leading to inefficiencies. Additionally, U.S. Space Command, as the global SATCOM manager, is unable to maintain situational awareness of which COMSATCOM resources DoD has access to, in which regions, and how much.

This fragmented approach is untenable for acquiring COMSATCOM services to support a hybrid architecture that integrates COMSATCOM and MILSATCOM capabilities. An enterprise approach is needed to acquire COMSATCOM services that can meet the requirements and attributes desired in a hybrid architecture (e.g., portability, interoperability, data-sharing for enterprise-level situational awareness).

CSCO is evaluating different acquisition strategies to consolidate COMSATCOM requirements for enterprise contracts while also supporting one-on-one contracts. We observed that accessing the commercial market in a more centralized way as a single buyer could result in greater influence on the market and better pricing. More importantly, enterprise contracts could provide better situational awareness and enable operational flexibility to dynamically shift resources among multiple providers, a capability that is needed to achieve assured connectivity in a contested environment.

That said, a consolidated approach and enterprise contracts (enabled by a working capital fund) might be met with resistance from user communities.[101] Interviewees noted that some user

---

[99] Warren Ferster, "Space Force to Centralize Commercial Satellite Procurement," *Government Satellite Report*, March 26, 2020.

[100] In 2020, CSCO had 100 different contracts and 30 more being negotiated at any given time (Ferster, 2020).

[101] In a working capital fund operating model, program expenses are charged to internal (in this case, DoD) customers, who must pay for them through their own funding sources. This ensures that supply and demand for a service are balanced. The Under Secretary of Defense (Comptroller) recently approved a working capital fund, the

communities (especially those that have access to funds to procure COMSATCOM) desire autonomy and have concerns that their specific mission needs might not be met or prioritized with consolidation. Interviewees also noted that some user communities perceive that the USSF should fund procurement of COMSATCOM services.

Lack of Maturity in Integrated Satellite Communications Operations Management

Realizing an integrated COMSATCOM-MILSATCOM architecture would also require an integrated approach to managing SATCOM operations. However, like the current approach to the procurement of COMSATCOM, the management of COMSATCOM operations (e.g., operational planning, SATCOM resource allocations) has been largely decentralized, leaving purchasing organizations to use their own internal processes and tools. There is no linkage between the users' operational management processes and tools for COMSATCOM and MILSATCOM operational management, which is centralized at U.S. Space Command (via Regional SATCOM Support Centers and Consolidated SATCOM Support Experts [C-SSEs]).[102]

As a result, no single DoD operational organization has a holistic view of COMSATCOM resources that are available to DoD users, who those users are, and the status of supporting commercial systems. Without enterprise-level situational awareness of both the COMSATCOM operations and the MILSATCOM operations, U.S. Space Command has limited ability to move users across multiple networks such that they maintain connectivity through all operating conditions.

Going forward, operations management of COMSATCOM will become centralized to align with the roles and responsibilities of U.S. Space Command as the global SATCOM manager in order to provide integrated SATCOM resources to the warfighter. This is especially critical in contested and degraded operational environments. To that end, U.S. Space Command assigned USSF, Headquarters Space Operations Command, to be the C-SSE for COMSATCOM and a direct support to the Combined Force Space Component Command in 2021.[103] The COMSATCOM C-SSE will serve as a counterpart to other C-SSEs for military systems (namely, wideband, narrowband, and protected) to enable an integrated SATCOM management framework. In this role, the COMSATCOM C-SSE will work with C-SSEs for military systems to provide operational support to COMSATCOM users.

As COMSATCOM C-SSE matures to improve integration of SATCOM management, much work remains. The visibility of COMSATCOM resources will need to be improved with the

---

"Enterprise Space Activity Group," under the USSF that would be managed by Space Systems Command to procure COMSATCOM ("The 'Color' of Space Money: Space Systems Command," *MilsatMagazine*, September 2022).

[102] SPI 3250.01A, 2020, provides additional details on roles and responsibilities of C-SSEs and Regional SATCOM Support Centers.

[103] Space Operations Command Operations Order 23-01, U.S. Space Command, December 2022.

capabilities from ESC-MC and with contracts that allow information-sharing. However, a policy question remains about U.S. Space Command's control over the allocation of COMSATCOM resources that purchasing organizations acquire.[104] As mentioned earlier, the purchasing organization manages its COMSATCOM resources, and contractual terms are unlikely to allow other DoD users to access those resources. Moreover, users might not want resource management to be centralized because they could perceive centralization as jeopardizing their ability to get the resources they want when needed.

Summary

COMSATCOM services provide DoD with a much-needed opportunity to enhance the capacity and resilience of the SATCOM enterprise. Our analysis revealed that COMSATCOM can help meet these needs without undue operational risk for a wide array of applications. Although each provider, satellite technology, or orbit may offer capabilities that are resilient against certain threats while vulnerable to others, operational risks can be significantly reduced if capabilities from multiple providers consisting of heterogeneous satellite technologies (e.g., narrow beams, intersatellite link, PTW) and multiple orbits are employed. Residual operational risks might be sufficiently acceptable for many missions requiring SATCOM, although we recognize that they might not be acceptable for certain critical missions (e.g., nuclear command, control, and communications).

The overall SATCOM market is likely to remain stable with a lot of potential for innovation and new services in the future, such as the emerging pLEO SATCOM market. The pLEO SATCOM market segment is worth monitoring closely while providers in that segment are facing greater risks in achieving business viability.

We assess that the greatest barrier to implementing a hybrid architecture to fully leverage diverse, heterogeneous capabilities offered by COMSATCOM is the integration of commercial capabilities into the DoD SATCOM ground segment and operations. Operational risks cannot be effectively mitigated without the capability for a user to access and roam between heterogeneous services from multiple providers, as well as between COMSATCOM and MILSATCOM. Acquisition strategies for and operations management of COMSATCOM also need to shift in order to realize an integrated commercial-military architecture. However, making these shifts is likely to be met with some resistance. Numerous stakeholders in the SATCOM enterprise need to coordinate and synchronize their investment decisions, and user communities need to be assured that their requirements will be met.

Drawing on these findings, we recommend that the DAF:

---

[104] According to SPI 3250.01A, 2020, Enclosure D, p. D-8, "With most COMSATCOM allocations, the resources acquired through COMSATCOM leasing are managed by the organization paying for the lease, which induces differences between the MILSATCOM and COMSATCOM allocation processes."

- leverage multiple SATCOM providers with heterogeneous technologies to mitigate operational risks by enforcing information assurance policies and collaborating with the user community and industry to improve awareness of operational risks
- establish a governance structure for synchronizing acquisition, operations, and requirements for integrating COMSATCOM and clarify roles and responsibilities among stakeholders
- develop a strategic plan for modernizing the SATCOM ground and terminal enterprises for unity of effort by
  - identifying key investments and outlining the phasing of those investments to achieve an increasing level of integration in a synchronized manner across space, ground, and user segments
  - utilizing the governance structure to coordinate and synchronize resourcing, capability development, and fielding of various ground segment elements and terminals undertaken by multiple stakeholder communities
- evaluate COMSATCOM requirements from an enterprise perspective and assess alternative acquisition strategies to meet enduring COMSATCOM requirements with enterprise contracts.

Chapter 5. Conclusions and Recommendations

The DAF is attempting to achieve four strategic goals in its engagement with the commercial space industry. It seeks to gain **access to innovation** that is being spurred by commercial demands. It wishes to maintain a **healthy space industrial base**. It seeks to gain **increased mission capabilities and resilience**. Finally, it hopes to acquire these capabilities **cost-effectively**. We examined the commercial PNT and COMSATCOM markets to identify risks to achieving those goals, and we offered recommendations specific to each market in Chapters Three and Four, respectively. In this chapter, we conclude by discussing overarching and generalizable findings and recommendations that the DAF should consider as it continues to engage with the growing commercial space industry.

Commercial Space Services Can Provide Operationally Useful Capabilities

In our analysis, we observed that commercial space services can provide increased mission capabilities and resilience to the DAF. Commercial capabilities can also provide additional capacity to scarce military capabilities. For example, DoD's SATCOM mission benefits from the commercial sector's contributions to overall capacity. Commercial providers also add resilience to the DAF's overall enterprise architecture when they provide capabilities independently. This is already the case in COMSATCOM; commercial PNT services might also add resilience if they can achieve faster time to first fix than GPS (assuming that operational risks can be overcome). Finally, commercial space services can provide new capabilities to DoD. For instance, if commercial PNT can develop capabilities that offer greater than 10-cm accuracy and greater signal strength, it can provide greater accuracy for precision strike and offer new blue force tracking options in restricted terrain that are not currently possible with GPS.

Our analysis also identified several operational risks that should be considered by the DAF. For example, our analysis suggests that commercial PNT services being developed today may be no more resistant to jamming than GPS. COMSATCOM is also vulnerable to jamming, cyberattacks, and ASAT and ground network attacks. However, our interviews with operators revealed that operational risks can be mitigated to some degree through tactics, techniques, and procedures and the judgment of commanders if they are thoroughly informed about the risks and their consequences. Additionally, each COMSATCOM provider, satellite technology, or orbit may offer capabilities that are vulnerable to certain threats but resilient to others. Thus, operational risks can be significantly reduced if capabilities from multiple providers consisting of multiple different satellite technologies (e.g., narrow beams, intersatellite link, PTW) and orbits are employed.

41

Commercial Space Services Present New Integration Risks

Gaining operationally useful capabilities from commercial space services will require the DAF to navigate partnerships with industry in new ways. Among other things, the DAF must articulate and implement hybrid architectures to achieve the operational benefits mentioned previously. DoD's SATCOM architecture is already headed toward a hybrid model to enable users to roam seamlessly between MILSATCOM and COMSATCOM as appropriate. In the PNT case, multiple signals might need to be accommodated and integrated into user equipment that uses only GPS today. The Space Warfighting Analysis Center is working to develop such hybrid architectures, but much work remains because commercial space service offerings continue to come to market. Similarly, leveraging the diverse COMSATCOM capabilities to reap the resilience benefits requires modernization of the SATCOM ground segment and terminals.

The prospect of new service offerings coming to market highlights the fact that commercial and government development timelines are not in sync. DoD capability development and acquisition timelines range from years to decades.[105] In contrast, commercial technology development timelines often last months. This misalignment has the potential to introduce new program management challenges as DoD tries to keep up with commercial development timelines.

Other integration risks that are peculiar to specific space services will also need to be addressed. For instance, commercial PNT services must contend with DoD restrictions on their use. As the DAF considers new commercial service offerings, it should prepare to deal with new integration risks stemming from the increased use of commercial services.

Market Maturity Will Have the Biggest Influence on the Ability of the Department of the Air Force to Meet All Its Commercial Goals

Although commercial space services offer the prospect of increased mission capabilities and resilience, achieving DoD's other goals of bolstering the industrial base and gaining access to innovation rely on the maturity of the commercial markets. Our examination of the commercial PNT and SATCOM markets' engagement with DoD suggests that commercial market maturity is the most significant factor in determining how well DoD can meet all four goals of leveraging commercial space services. We consider a market to be mature if firms participating in the market can generate sustainable revenues from service offerings to a diversified customer base.

The COMSATCOM market described in Chapter Four is an example of a mature market. Commercial demand is known, as are the resulting revenue streams. This provides space for investors and SATCOM providers to explore new technologies, such as pLEO constellations and

---

[105] Jonathan P. Wong, Obaid Younossi, Christine Kistler LaCoste, Philip S. Anton, Alan J. Vick, Guy Weichenberg, and Thomas C. Whitmore, *Improving Defense Acquisition: Insights from Three Decades of RAND Research*, RAND Corporation, RR-A1670-1, 2022.

multiservice offerings, without having to "bet the business" on immediate success. There is enough financial cushion to support technology development and predictable demand to make that development worthwhile.

The customer base must consist of DoD *and* others to minimize the likelihood of DoD becoming a dominant or sole customer. If DoD becomes the dominant or sole customer, the industrial base will be limited, and market or constituent firms will not be as incentivized to innovate.

Investing in Capabilities from an Immature Market Risks the Department of the Air Force's Long-Term Goals

The DAF needs to be aware of the implications of investing in immature markets and measure its investment strategy against potential risks. This is not to say that DoD should not acquire commercial space services from immature markets in all circumstances. Consider the prospects for space-based commercial PNT. We observed that the long-term commercial viability of this market is limited, even under the most-optimistic financial circumstances. But DoD may judge that the immediate benefit of the capability to augment the GPS system is worth considering. Such a judgment is outside the scope of this research, but there are clearly some DoD users who are willing to consider it.[106]

If DoD decides to make such an investment to fulfill an enduring capability need, it should be prepared to make trade-offs between its long-term goals. Consider again the space-based commercial PNT case. If the capability were valuable enough for DoD to fund it as a program of record (or part of one), two outcomes might result. One would be that DoD's investment might drive the cost of this service down such that the service expands the addressable commercial market to the point that a stable revenue stream is possible for several firms. This is the most favorable outcome.

However, DoD's investment might not jumpstart further investments. In that case, DoD may be left as the dominant or sole customer of space-based commercial PNT services. The value of DoD's investment may be limited to the immediate capability that the service provides. Cost-effectiveness may erode over time without competitive market forces. The market itself will likely erode as well, as was the case with DoD's attempt to commercialize remote sensing that began in the 1980s.[107] This will limit the industrial base to a few commercial providers, or even one provider. Further innovation will be limited to the remaining firms attempting to fulfill DoD requirements, and there will be a lack of commercial demand to attract new ideas. Either

---

[106] One example is the U.S. Army's recent focus on assured PNT. See U.S. Army Acquisition Support Center, "Assured-Positioning, Navigation and Timing (A-PNT)—Dismounted," webpage, undated.

[107] See Appendix A for details on the remote-sensing market.

outcome in such a limited market scenario will likely prevent DoD from meeting all of its strategic goals for leveraging commercial space capabilities.

Recommendations for the Department of the Air Force

As the DAF considers the risks and benefits of leveraging any given commercial space service offering, it should consider whether achieving all four strategic goals is possible or whether it must prioritize a smaller set. The DAF should understand what its desired end state looks like before it adopts a risk-tolerant or risk-averse approach to each commercial service offering.

Considerations for a Risk-Tolerant Approach

When one or more of the following conditions prevail, the DAF should be more willing to leverage commercial space services as the primary means of fulfilling a capability need, for sensitive missions, or for other situations in which the investment cannot be easily wound down:

- Commercial demand for similar services exists and supports sustainable revenue streams for firms participating in the market.
- Market risk can be spread across stakeholders, especially others within the U.S. government.
- The commercial service provides a new or urgently needed capability that cannot be met otherwise.
- The integration of the service can be reversed or unwound through open architecture standards or other means.

Considerations for a Risk-Averse Approach

When it wishes to take a risk-averse approach, the DAF should adopt a wait-and-see approach by commissioning and supporting prototypes and other pilot programs through defense innovation organizations, such as the Defense Innovation Unit, AFWERX, and SpaceWERX, that administer Small Business Innovation Research and rapid-prototype programs. Doing so would enable DoD to monitor and support the development of these services while allowing commercial demand to build up. This would minimize the chance that DoD might not meet all four of its goals for leveraging commercial space services. The DAF should be more risk averse when one or more of the following conditions prevail:

- There is unclear or limited commercial start-up funding or sustainable revenue.
- Only a few firms (or one firm) are working in the space.
- The operational value of the commercial service is not yet clear or is not likely to offer an appreciable advantage over the traditionally acquired capability.

No matter what posture the DAF adopts toward commercial space service offerings, it should make three key refinements to the way it acquires commercial space services.

First, the DAF should **invest in greater market intelligence capabilities**. We observed a dichotomy in our stakeholder interactions. Some DAF stakeholders of a given capability were very bullish on the prospects of a given commercial service offering, particularly those whose work exposes them to commercial providers and their capabilities. Others with little to no interaction with commercial providers expressed the viewpoint that the commercial offerings were inappropriate to meet DAF needs. Our own engagement with commercial firms and our analysis suggest that the reality is likely somewhere in the middle. To that end, the DAF should continue to build its market intelligence capabilities so that it understands not only the technical capabilities being developed but also contextual factors, such as financial viability and market dynamics, that will inform any decision to leverage a commercial service.

Second, the DAF should **increase the sophistication of its contracting capabilities** to be more adept at negotiating contracted services. Previous RAND research highlights the increased challenge of tailoring contract clauses for nontraditional cases, such as other transactions.[108] Stakeholder interviews suggest that negotiating agreements and contracts for commercial space services will be more challenging than contracting with traditional defense suppliers. It is particularly important to ensure that service contracts align with the DAF's chosen risk posture and prioritization of strategic goals for a given capability.

Finally, related to contracting, the DAF should work to **build flexible resourcing options** to responsively negotiate service contracts. Traditional defense budgeting requires two years from the time a requirement is articulated to the time funds are available. Reallocating those funds after they are appropriated requires the DAF to go through congressional reprogramming requests that can be time-consuming. These are unacceptably slow processes. Conversely, tactical and operational commands using operations and maintenance funds to acquire services make it difficult for the DAF to have a true picture of the extent to which commercial space services are being used. A central, flexible funding source, such as a working capital fund that funds service contracts centrally via user funding sources, can help address the visibility and responsiveness problems.

Continued Growth of Commercial Space

Activities in the commercial space market will continue regardless of whether the DAF makes significant investments. However, the conditions that we have articulated for adopting a

---

[108] Lauren A. Mayer, Mark V. Arena, Frank Camm, Jonathan P. Wong, Gabriel Lesnick, Sarah Lovell, Edward Fernandez, Phillip Carter, and Gordon T. Lee, *Prototyping Using Other Transactions: Case Studies for the Acquisition Community*, RAND Corporation, RR-4417-AF, 2020. An *other transaction* is a type of contract used by DoD.

more risk-tolerant or risk-averse approach to any given commercial service offering can inform more-intentional and more-deliberate decisionmaking and allow the DAF to benefit from commercial offerings.

Appendix A. Remote-Sensing Case Study

The early remote-sensing commercial market, which existed from the 1990s to the 2010s, provides an insightful look into the development of the commercial space industry and the roles that the government and DoD played in that market. The development of that market has similar characteristics to how today's commercial space industry is developing and growing.

Market Creation and Development

A series of policies, including the Land Remote-Sensing Commercialization Act of 1984,[109] the Land Remote Sensing Policy Act of 1992,[110] and numerous National Security Presidential Directives,[111] ushered in the creation of the U.S. remote-sensing market. A key driver behind these policies was a concern that the existing regulatory landscape hindered the United States' competitiveness in remote-sensing technologies.[112]

Prior to the passage of the Land Remote Sensing Policy Act of 1992, no commercial firms had licenses to operate privately owned remote-sensing satellites. After the bill's passage, more than a dozen firms quickly entered the market and obtained licenses.[113] These firms sought to capitalize on a new and developing market. Estimates for the growth of the commercial remote-sensing market from 1992 to the turn of the century varied wildly; predictions ranged from an annual growth factor of 5 to a factor of 50 over a ten-year period.[114]

Market Consolidation and Government Power

At the time, firms faced several challenges, including financing, firm longevity, affordable insurance, spectrum allocation, orbital slot allocation, technical risk, and challenges operating in

---

[109] Public Law 98-365, Land Remote-Sensing Commercialization Act of 1984, July 17, 1984.

[110] Public Law 102-555, Land Remote Sensing Policy Act of 1992, October 28, 1992.

[111] National Security Presidential Directive 1, "Organization of the National Security Council System," White House, February 13, 2001; National Security Presidential Directive 2, "Improving Military Quality of Life," White House, February 15, 2001; National Security Presidential Directive 3, "Defense Strategy, Force Structure, and Procurement," White House, February 15, 2001.

[112] Kennedy et al., 2019.

[113] National Oceanic and Atmospheric Administration Satellite and Information Service, "Licensing of Commercial Remote Sensing Satellite Systems," webpage, last updated April 20, 2003.

[114] C. Bryan Gabbard, Kevin M. O'Connell, George S. Park, and Peter J. E. Stan, *Emerging Markets of the Information Age: A Case Study in Remote Sensing Data and Technology*, RAND Corporation, DB-176-CIRA, 1996.

the existing policy landscape (i.e., challenges with foreign sales).[115] Today, firms face many of these same challenges.

Despite aggressive predictions for growth in the remote-sensing market in the early 1990s, the market saw mass consolidation; from the initial dozen-plus firms, only one commercial provider remained by the end of 2012. Because of a series of mergers and acquisitions, the market was served by a sole firm for several years before a new generation of remote-sensing firms began to enter the market.

A key force in driving consolidation during this time was the government's monopsonist behavior. The awarding of significant government contracts, such as ClearView, NextView, and EnhancedView, was often followed by the losing firm being acquired or merging with the contract awardee. These contracts represented a core revenue stream for firms, and losing this revenue was cited as the key reason for a merger or acquisition.[116] This pattern highlights the outsized influence a large customer like the government carries in developing markets.

Lessons Learned and Parallels to Today

Today's commercial space market, which is not limited to remote sensing, shares key similarities with the market of the early 1990s to 2010s. There is tremendous interest in commercial space, and key policy actions are being taken to enable the growth of the commercial space market. Furthermore, there is an ever-increasing number of firms involved in the industry. We outline some key lessons here.

First, policy can be an effective tool in stimulating and creating new markets, just as a lack of policy can stunt the development of new markets. Prior to 1992, there were no licensed firms to commercially operate private remote-sensing space systems, but the market grew quickly after the passage of key bills. However, in the case of synthetic aperture radar systems, the United States was slow to adopt and create a supportive policy environment for firms pursuing these systems and effectively ceded the commercial market to foreign firms. Only recently has the regulatory landscape evolved to enable U.S. firms to make up lost ground.[117]

Second, in developing markets that are limited in size, a large government customer can exert significant power over the outcomes of firms in that market. The government's award of the ClearView, NextView, and EnhancedView contracts played a significant role in the subsequent mergers and acquisitions of firms that were not awarded these contracts. However, these contract awards simultaneously demonstrate how government support can provide firms with financial

---

[115] Gabbard et al., 1996.

[116] GeoEye, Inc., Form 425, Securities and Exchange Commission filing, November 1, 2012.

[117] Sandra Erwin, "Report: Lack of Government Investment Hurting U.S. Geospatial Intelligence Industry," *SpaceNews*, October 29, 2020.

stability. In today's commercial space market, the government must simultaneously play the roles of customer, regulator, and competitor.

These lessons informed the conceptual framework presented in Chapter 2 and the interpretation of findings in Chapters 3 and 4 that are the basis of our policy recommendations. We paid particular attention to the relative power between commercial firms and government customers and its effects on relevant market dynamics.[118]

---

[118] It is worth noting that new commercial remote-sensing opportunities were starting to emerge as this research was being completed. See, for instance, Sandra Erwin, "NRO Signs Agreements with Six Commercial Providers of Space-Based RF Data," *SpaceNews*, September 28, 2022d.

Appendix B. Complementing an Inertial Measurement Unit with Commercial Positioning, Navigation, and Timing

Although the focus of this research was on space-based commercial services, our exploration of the PNT case revealed several non–space-based PNT capabilities that could be used in conjunction with space-based solutions to increase accuracy. We examine one such capability in this appendix.

We consider an approach in which commercial PNT is used as a third source to improve an onboard PNT system. We assume a design in which an onboard IMU is coupled with a GPS receiver in an INS to provide position and navigation information if the GPS signal is not available. The IMU is assumed to be regularly disciplined by GPS (when that signal is available) and becomes the primary provider of the needed measurement when GPS is denied.[119]

Once the GPS signal is unavailable, the INS performance, which relies on the IMU, degrades with time. However, if the INS can also receive a commercial PNT signal, it can then replace the GPS signal to calibrate the onboard IMU to reduce the measurement error. Figure B.1 shows an approximate representation of the position error with and without the GPS calibration (which represents the commercial capability).[120]

We estimated the potential value provided by an INS coupled with both a GPS and a commercial receiver. The analysis used GPS as a surrogate to the commercial PNT system; however, the performance would vary depending on several factors, including the quality of the PNT system provided by the specific commercial provider being used, the PNT outage update time, and the maneuvers to which the INS would be subjected.

As the figure shows, the calibration provides about a 70-percent improvement in the position accuracy measurement. We also assumed that the IMU is provided with algorithms to detect spoofing by comparing the instantaneous onboard measurement with the received signal; a large variance in the comparison would indicate potential signal spoofing. This approach would significantly limit the vulnerability of the INS to spoofing.

---

[119] *Disciplined by GPS* is a term of art that implies that the INS is calibrated by the GPS signal.

[120] The IMU performance improvement level is based on example results from a published higher-fidelity analysis (E. Mooij and Q. P. Chu, "IMU/GPS Integrated Navigation System for a Winged Re-Entry Vehicle," paper presented at American Institute of Aeronautics and Astronautics Guidance, Navigation, and Control Conference and Exhibit, Montreal, August 6–9, 2001) and is intended only to illustrate the potential value of integrating an off-board PNT signal; a detailed analysis using a Kalman filter (or an equivalent) is needed to estimate the actual performance of an INS for different operational conditions.

Figure B.1. Potential Inertial Measurement Unit Performance Improvement

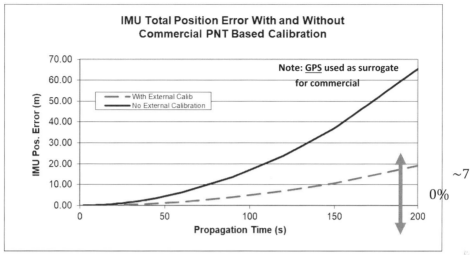

SOURCE: Example results are from Mooij and Chu, 2001.
NOTE: Pos. = position. Analysis is based on a Honeywell HG9900 IMU and a 0.2-g acceleration condition.

Appendix C. Risk Landscape for Proliferated Low Earth Orbit

The ambition to develop LEO SATCOM constellations is not new, but the risks and landscape have changed over time. In the 1990s, Globalstar, Odyssey, Teledesic, and Iridium, among others, pioneered and competed for the LEO market. Only Iridium and Globalstar—both of which continue to offer services today—survived, even as terrestrial cell phone providers increased bandwidth. High costs and limited demand kept the market too small for the competition.

Today's demand landscape is different, as the state of technology has generally advanced and demand for bandwidth has soared across every sector. This increased demand generally creates an opportunity for growth in the SATCOM market, and the need for more internet communications is well beyond current supply. But growth in demand is not sufficient to ensure a firm's viability. In the process of pursuing a place in the pLEO market, firms need to secure spectrum access, orbital deployment authority, and the authority to operate while in space and then ensure financial sustainability thereafter—two topics discussed in this appendix.

Regulatory Requirements

Two major regulatory factors contribute to risk for firms entering the pLEO market. The first, securing access to the spectrum and orbital resources, is considered by many to be the biggest challenge to pursuing pLEO. Spectra and orbits are finite, so once a regulatory agency—such as the FCC, the International Telecommunication Union, or an FCC-equivalent agency in another country—allocates those resources, they are locked until the winning firm fails to meet the criteria to operate in the market or reallocates its spectral and orbital rights. For example, in the V-band spectrum, more than 94,000 satellites have been proposed to the FCC in a rush to secure access to the spectrum and orbits.[121]

When incumbent firms with existing licenses or an ability to navigate the regulatory space are favored in the allocation of the spectrum regardless of whether they can deliver services, it is more difficult for new firms to enter the market. To ensure that licenses are used, the FCC does have a caveat in its allocations that a firm must launch at least half of its proposed satellites by the sixth year of a license and all of its proposed satellites by the ninth year. However, the FCC does not necessarily consider the soundness or feasibility of a business plan in its rulemaking. If a firm occupies a part of the spectrum or orbits and fails to deliver, this could lead to an inefficient use of resources. At the same time, a firm's ability to innovate can be stymied by the potentially time-consuming process of obtaining a license from the FCC.

---

[121] Messier, 2021.

The second regulatory factor that contributes to risk is the impact of the FCC's and similar regulatory organizations' decisionmaking timelines on schedules for delivery of services. Furthermore, when the spectrum has been allocated, the decision has not always been final. For example, the FCC reallocated a portion of the C-band spectrum to the 5G industry in 2020.[122] These uncertainties add to the financial risks that firms face as they navigate the process of obtaining access to the spectrum.

Financial Hurdles

Participating in the pLEO market requires a great deal of up-front capital to provide a high-cost service with a long-term profit horizon. The technology itself is risky and expensive. In addition, high costs associated with launch, spacecraft manufacturing, ground equipment, and user equipment remain a problem.

Even though demand is high, customers are not willing or able to pay above a certain price. For example, under a funding program called the *Rural Digital Opportunity Fund*, the FCC first approved funding in December 2020 but then reversed its decision in August 2022, when it deemed the planned product cost-prohibitive.[123]

While firms are developing SATCOM as a service, subscription models have their own trade-offs. Subscription models allow customers flexibility to choose providers and to switch providers with little to no friction as better or cheaper competing services become available, resulting in risk of customer attrition.

With a subscription model, legacy firms eventually might become vulnerable because of their higher overhead costs associated with deeper innovation, higher sunk costs, vendor lock-in, more internal bureaucracy, and a stronger bias toward the status quo. Younger firms specializing in unprecedented technologies might have the agility and flexibility to produce the needed services, but they might at the same time be constrained by lack of capital and reputation, which poses financial risks. OneWeb and Inmarsat were combined with Eutelsat and ViaSat, respectively, in 2022, after several financial setbacks.[124] While the smaller, younger companies proved to be financially inviable on their own, more-mature companies might be able to more effectively mitigate their respective risks—by absorbing the younger companies.

---

[122] Rachel Jewett, "FCC Cancels Starlink Funding for Rural Broadband Program," Via Satellite, August 10, 2022.

[123] Jewett, 2022.

[124] Inmarsat, "Viasat and Inmarsat to Combine, Creating a New Leading Global Communications Innovator," press release, November 8, 2021; OneWeb, 2022b.

Appendix D. Interview Protocols

During this research, we conducted semi-structured interviews with more than 30 stakeholders from a variety of organizations. Interviews were conducted in person and via Microsoft Teams and followed informed consent procedures outlined by the RAND Corporation's Human Subjects Protection Committee guidelines.

Semi-structured interviews give the interviewer and interviewee wide latitude to discuss issues that are related to the topic at hand but that might not be covered directly in the protocol. Therefore, these protocols served only as guides and not as strict scripts to adhere to.

Interview Protocol for Positioning, Navigation, and Timing

1. Can you describe your role and responsibilities as they pertain to PNT?
2. What are the different ways in which commercial capabilities and services can be leveraged for PNT?
3. What commercial services can be used for position/navigation versus timing?
4. Which military applications are strongly dependent on PNT?
5. What are the risks from your perspective in leveraging commercial services for PNT?
6. What are the drivers of these risks (programmatic, technical, business)?
7. What are the impacts of these risks to PNT mission effectiveness?
8. Are there other, larger impacts that should be considered when determining how to leverage commercial PNT?
9. How do risks differ depending on how commercial services are used for position/navigation versus timing?
10. What are the current approaches for mitigating/addressing these risks?
11. Are these approaches in the form of investments, modifying the use case, or something else?
12. How do mitigation options differ depending on how commercial services are used for position/navigation versus timing?
13. What acquisition models exist for leveraging commercial PNT?
14. Do the risks differ by acquisition approach?
15. Is there anything else you would like to share with us?
15. Are there other people that we should talk to or other documents you can provide or think we should read?

Interview Protocol for Commercial Satellite Communications

1. What are your organization's roles and responsibilities as they pertain to SATCOM?
2. What are your background and responsibilities in this organization?
3. How do commercial SATCOM services support your organization's mission and activities?
4. How do you integrate military and commercial capabilities?

5. What new or emerging commercial services is your organization considering?
6. What are the risks from your perspective in leveraging commercial services for SATCOM?
7. What are the drivers of these risks (programmatic, technical, business)?
8. What are the impacts of these risks to PNT mission effectiveness?
9. Are there other, larger impacts that should be considered when determining how to leverage commercial SATCOM?
10. What are the current approaches for mitigating/addressing these risks?
11. What acquisition models do you use to leverage commercial SATCOM services?
12. What acquisition challenges exist for acquiring commercial SATCOM services?
13. How can those challenges be addressed?
14. Is there anything else you would like to share with us?
15. Are there other people that we should talk to or other documents you can provide or think we should read?

Interview Protocol for the Acquisition Process

1. What are your organization's roles and responsibilities as they pertain to leveraging commercial space capabilities?
2. What are your background and responsibilities in this organization?
3. What is your organization's approach to acquiring commercial capabilities for the DoD?
4. How does your organization collaborate with other DoD entities to acquire commercial capabilities?
5. What involvement do you have with the acquisition of commercial PNT or SATCOM capabilities?
6. What are the biggest impediments to leveraging commercial capabilities?
7. What is your perspective on managing the fundamental risks associated with leveraging commercial capabilities over military ones?
8. In your opinion, what changes to DAF policies or practices are needed to better support the acquisition of commercial capabilities?
9. Is there anything else you would like to share with us?
10. Are there other people that we should talk to or other documents you can provide or think we should read?

Appendix E. Affiliations of Experts and Stakeholders Interviewed

Table E.1 lists the organizational affiliations and areas of expertise of the subject-matter experts we interviewed.

Table E.1. Interviewees' Affiliations and Areas of Expertise

Area of Expertise	Organization
SATCOM	U.S. Indo-Pacific Command, Operations Directorate, Space and Integrated Air and Missile Defense
	U.S. Space Command Joint Integrated Space Team
	U.S. Indo-Pacific Command, Command, Control, Communications, and Cyber Directorate (J6) and component commands
	Space Warfighting Analysis Center
	U.S. Army Satellite Operations Brigade
	U.S. Space Command Strategy, Plans and Policy Directorate (J5)
	U.S. Space Command Force Structure, Resources, and Assessment Directorate (J8)
	Space Operations Command SATCOM Mission Area Team
	Chief of Space Operations for Strategy and Resources
	SAF/SQ SATCOM Team
	DoD Chief Information Officer
PNT	Space Warfighting Analysis Center
	Space Systems Command
	SAF/SQX PNT Team
	Space Operations Command PNT Mission Area Team
	Space Systems Command PNT Futures
	DoD Chief Information Officer
	Army Futures Command Assured PNT Cross Functional Team
	Department of Commerce Commercial Space Office
	Xona Space Systems
	TrustPoint
	Satelles, Inc.
	Johns Hopkins University Applied Physics Laboratory
	Chief of Space Operations for Strategy and Resources
	U.S. Indo-Pacific Command, Operations Directorate, Space and Integrated Air and Missile Defense

Area of Expertise	Organization
Acquisition	AFWERX and SpaceWERX
	Space Rapid Capabilities Office
	SAF/SQ
	National Reconnaissance Office Commercial Systems Program Office
Other	DoD Chief Technology Information Officer
	Commercial Integration Cell

Abbreviations

ASAT	anti-satellite
COMSATCOM	commercial satellite communications
CSCO	Commercial Satellite Communications Office
C-SSE	Consolidated SATCOM Support Expert
DA	direct ascent
DAF	Department of the Air Force
DE	directed energy
DoD	Department of Defense
ESC-MC	Enterprise Satellite Communications Management and Control
EUSPA	European Union Agency for the Space Programme
EW	electronic warfare
FCC	Federal Communications Commission
FSS	fixed-satellite services
GEO	geosynchronous orbit
GNSS	Global Navigation Satellite System
GNSS-RO	Global Navigation Satellite System Radio Occultation
HTS	high-throughput satellite
IMU	inertial measurement unit
INS	integrated navigation system, *or* inertial navigation system
KKV	kinetic kill vehicle
LEO	low earth orbit
MEO	medium earth orbit
MGUE	Military GPS User Equipment
MILSATCOM	military satellite communications
MSS	mobile-satellite services
pLEO	proliferated low earth orbit
PNT	positioning, navigation, and timing
pntOS	PNT Operating System
PTW	protected tactical waveform
SAF/SQ	Assistant Secretary of the Air Force for Space Acquisition and Integration
SATCOM	satellite communications

SPI	U.S. Space Command Instruction
STL	Satellite Time and Location
USSF	U.S. Space Force

References

Alkire, Brien, Jonathan Fujiwara, Moon Kim, Yool Kim, George Nacouzi, Colby Steiner, and James Williams, *Leveraging Commercial Space Internet Services for Air Missions*, RAND Corporation, 2020, Not available to the general public.

Baghal, Lisa A., *SATCOM Supply Versus Demand and the Impact on Remotely Piloted Aircraft ISR*, Air Command and Staff College, Air University, March 2016.

Bonds, Tim, Michael G. Mattock, Thomas Hamilton, Carl Rhodes, Michael Scheiern, Philip M. Feldman, David R. Frelinger, and Robert Uy, *Employing Commercial Satellite Communications: Wideband Investment Options for the Department of Defense*, RAND Corporation, MR-1192-AF, 2000. As of August 29, 2022: https://www.rand.org/pubs/monograph_reports/MR1192.html

BryceTech, *State of the Satellite Industry Report*, Satellite Industry Association, June 2021a.

BryceTech, "Smallsats by the Numbers: 2021," presentation slides, August 13, 2021b.

BryceTech, *State of the Satellite Industry Report*, Satellite Industry Association, June 2022.

Calvelli, Frank, and David D. Thompson, "Fiscal Year 2023 Priorities and Posture of the U.S. Space Force," statement presented to the Subcommittee on Strategic Forces, U.S. Senate, May 11, 2022.

Chairman of the Joint Chiefs of Staff Instruction 6130.01G, *2019 Chairman of the Joint Chiefs of Staff Master Positioning, Navigation, and Timing Plan*, Joint Chiefs of Staff, 2019.

Christensen, Ian, "Trends and Developments in Commercial Space Situational Awareness," presentation slides for seminar, Secure World Foundation, April 7, 2021.

"The 'Color' of Space Money: Space Systems Command," *MilsatMagazine*, September 2022.

Commercial Satellite Communications Office, *Fiscal Year 21 Commercial Satellite Communications Expenditures and Usage Report*, Space Systems Command, U.S. Space Force, 2021, Not available to the general public.

Congressional Budget Office, *The Global Positioning System for Military Users: Current Modernization Plans and Alternatives*, Pub. No. 4192, October 2011.

Davenport, Christian, "Disrupted by SpaceX, ULA Was in 'Serious Trouble.' Now It's on the Road Back," *Washington Post*, June 22, 2022.

Department of Defense Instruction 8420.02, *DoD Satellite Communications*, U.S. Department of Defense, November 25, 2020.

DoD—*See* U.S. Department of Defense.

Duke, Hannah, "On-Orbit Servicing," Center for Strategic and International Studies, September 16, 2021.

Erwin, Sandra, "Air Force Seeking Commercial Technologies for Cislunar Space Operations," *SpaceNews*, December 12, 2019.

Erwin, Sandra, "Report: Lack of Government Investment Hurting U.S. Geospatial Intelligence Industry," *SpaceNews*, October 29, 2020.

Erwin, Sandra, "Space Force Signals Demand for Commercial Weather Data, but Will the Industry Deliver?" *SpaceNews*, January 17, 2022a.

Erwin, Sandra, "Space Force Looking to Ease Barriers to Entry for Commercial Companies," *SpaceNews*, April 4, 2022b.

Erwin, Sandra, "Military Buyers Challenged to Stay Up on the Latest Commercial Space Innovations," *SpaceNews*, May 18, 2022c.

Erwin, Sandra, "NRO Signs Agreements with Six Commercial Providers of Space-Based RF Data," *SpaceNews*, September 28, 2022d.

Euroconsult, "High Throughput Satellites Poised to Become Leading Commercial Growers in Space Infrastructure, with Wholesale Capacity Revenues Projected to Top $100b by 2030," press release, March 23, 2022.

European Union Agency for the Space Programme, *EUSPA EO and GNSS Market Report*, No. 1, 2022.

EUSPA—*See* European Union Agency for the Space Programme.

Eutelsat, "Combining Satellite and OTT Delivery," brochure, undated.

Eutelsat Communications Group, *Consolidated Financial Statements as of 30 June 2020*, undated-a.

Eutelsat Communications Group, *Consolidated Financial Statements as of 30 June 2021*, undated-b.

Eutelsat Communications Group, *Consolidated Financial Statements as of 30 June 2022*, undated-c.

Ferster, Warren, "Space Force to Centralize Commercial Satellite Procurement," *Government Satellite Report*, March 26, 2020.

Fishman, Arthur, and Rafael Rob, "Product Innovation by a Durable-Good Monopoly," *RAND Journal of Economics*, Vol. 31, No. 2, Summer 2000.

Foust, Jeff, "OneWeb Continues to Study Offering Navigation Services," *SpaceNews*, April 8, 2021a.

Foust, Jeff, "Launch Companies Optimistic About Future Demand," *SpaceNews*, September 9, 2021b.

Foust, Jeff, "Satellite Operators Need More Accurate SSA Data," *SpaceNews*, September 16, 2021c.

Foust, Jeff, "GEO Operators Say They Can Compete Against LEO Systems on Cost," *SpaceNews*, March 22, 2022.

Gabbard, C. Bryan, Kevin M. O'Connell, George S. Park, and Peter J. E. Stan, *Emerging Markets of the Information Age: A Case Study in Remote Sensing Data and Technology*, RAND Corporation, DB-176-CIRA, 1996. As of August 26, 2022: https://www.rand.org/pubs/documented_briefings/DB176.html

GeoEye, Inc., Form 425, Securities and Exchange Commission filing, November 1, 2012. As of September 1, 2022: https://www.sec.gov/Archives/edgar/data/0001040570/000119312512449233/ d433165d425.htm

GPS.gov, "GPS Accuracy," webpage, last modified March 3, 2022. As of September 28, 2022: https://www.gps.gov/systems/gps/performance/accuracy/

Gutierrez, Peter, "Fleshing Out the Leo PNT Landscape," *Inside GNSS*, March 14, 2022.

Harris, Mark, "SpaceX's Starlink Satellites Could Make US Army Navigation Hard to Jam," *MIT Technology Review*, September 28, 2020.

Hitchens, Theresa, "Space Force to Focus SATCOM Management on JADC2 Needs: Exclusive," *Breaking Defense*, December 18, 2020.

Hitchens, Theresa, "Antennas: The Hard Physics Challenge for Space Force 'Hybrid' SATCOM Plan," *Breaking Defense*, April 15, 2022.

Inmarsat, "Viasat and Inmarsat to Combine, Creating a New Leading Global Communications Innovator," press release, November 8, 2021.

Inmarsat Global Limited, *Annual Report and Financial Statements for the Year Ended 31 December 2020*, registered number 3675885, September 29, 2021.

Inmarsat Global Limited, *Annual Report and Financial Statements for the Year Ended 31 December 2021*, registered number 3675885, September 30, 2022.

Intelsat, *Annual Report Pursuant to Section 13 or 15(d) of the Securities Exchange Act of 1934 for the Fiscal Year Ended December 31, 2019*, U.S. Securities and Exchange Commission, commission file 001-35878, February 2020.

Intelsat, *Annual Report Pursuant to Section 13 or 15(d) of the Securities Exchange Act of 1934 for the Fiscal Year Ended December 31, 2020*, U.S. Securities and Exchange Commission, commission file 001-35878, March 2021.

Iridium Communications, *Annual Report 2019*, 2020.

Iridium Communications, *Annual Report 2020*, 2021.

Iridium Communications, *Annual Report 2021*, 2022.

Jewett, Rachel, "FCC Cancels Starlink Funding for Rural Broadband Program," Via Satellite, August 10, 2022.

Joint Publication 3-14, *Space Operations*, Joint Chiefs of Staff, April 10, 2018, change 1, October 26, 2020.

Keller, John, "Northrop Grumman to Advance Position, Navigation, and Timing (PNT) for Secure Communications Satellites," Military & Aerospace Electronics, May 6, 2021.

Kennedy, Michael, Yool Kim, Brien Alkire, Benjamin M. Miller, Stephanie Young, Therese Marie Jones, and Matthew Sargent, *Analysis of Commercial Space Capabilities: Leveraging New Space to Increase Resilience of DoD Space—Business Viability Analysis*, RAND Corporation, 2019, Not available to the general public.

Kim, Yool, George Nacouzi, Mary Lee, Brian Dolan, Krista Romita Grocholski, Emmi Yonekura, Moon Kim, Thomas Light, and Raza Khan, *Leveraging Commercial Space Capabilities to Enhance the Space Architecture of the U.S. Department of Defense*, RAND Corporation, 2022, Not available to the general public.

Kim, Yool, Ellen Pint, David Galvan, Meagan Smith, Therese Marie Jones, and William Shelton, *How Can DoD Better Leverage Commercial Space Capabilities? Understanding Business Processes and Practices in the Commercial Satellite Service Industry*, RAND Corporation, 2016, Not available to the general public.

Lardinois, Frederic, "Weather Platform ClimaCell Is Now Tomorrow.io and Raises $77M," Tech Crunch, March 30, 2021.

Mason, Richard, James Bonomo, Tim Conley, Ryan Consaul, David R. Frelinger, David A. Galvan, Dahlia Anne Goldfeld, Scott A. Grossman, Brian A. Jackson, Michael Kennedy, Vernon R. Koym, Jason Mastbaum, Thao Liz Nguyen, Jenny Oberholtzer, Ellen M. Pint, Parousia Rockstroh, Melissa Shostak, Karlyn D. Stanley, Anne Stickells, Michael J. D. Vermeer, and Stephen M. Worman, *Analyzing a More Resilient National Positioning, Navigation, and Timing Capability*, Homeland Security Operational Analysis Center operated by the RAND Corporation, RR-2970-DHS, 2021. As of September 19, 2022: https://www.rand.org/pubs/research_reports/RR2970.html

Mayer, Lauren A., Mark V. Arena, Frank Camm, Jonathan P. Wong, Gabriel Lesnick, Sarah Lovell, Edward Fernandez, Phillip Carter, and Gordon T. Lee, *Prototyping Using Other Transactions: Case Studies for the Acquisition Community*, RAND Corporation, RR-4417-AF, 2020. As of August 25, 2022:
https://www.rand.org/pubs/research_reports/RR4417.html

Messier, Doug, "Planned Comsat Constellations Now Exceed 94,000 Satellites," Parabolic Arc, November 8, 2021.

Military Satellite Communication Analysis, Systems Integration and Test, and Engineering Services, "Protected Tactical Enterprise Service (PTES) - Protected Tactical Waveform (PTW) Interface Control Document (ICD)," Air Force Space and Missile Systems Center, MILSATCOM Systems Directorate (SMC/MC), March 9, 2018, Not available to the general public.

Miller, Amanda, "Jam-Resistant Waveform for Safer Battlefield Communications to Get First In-Space Test," *Air Force Magazine*, March 23, 2022.

Mooij, E., and Q. P. Chu, "IMU/GPS Integrated Navigation System for a Winged Re-Entry Vehicle," paper presented at American Institute of Aeronautics and Astronautics Guidance, Navigation, and Control Conference and Exhibit, Montreal, August 6–9, 2001.

Morin, Jamie, and Robert S. Wilson, *Leveraging Commercial Space for National Security*, Center for Space Policy and Strategy, Aerospace Corporation, November 2020.

Nacouzi, George, Yool Kim, Jake McKeon, Alex Sedlack, Karishma R. Mehta, Mel Eisman, and Myron Hura, *Acquisition of Future Proliferated Low Earth Orbit Programs: Potential Challenges in Acquiring and Sustaining Mega Constellations*, RAND Corporation, 2020, Not available to the general public.

National Oceanic and Atmospheric Administration Satellite and Information Service, "Licensing of Commercial Remote Sensing Satellite Systems," webpage, last updated April 20, 2003. As of September 1, 2022:
https://web.archive.org/web/20041206081614/http://www.licensing.noaa.gov/licensees.html

National Security Presidential Directive 1, "Organization of the National Security Council System," White House, February 13, 2001.

National Security Presidential Directive 2, "Improving Military Quality of Life," White House, February 15, 2001.

National Security Presidential Directive 3, "Defense Strategy, Force Structure, and Procurement," White House, February 15, 2001.

"A New Class of Protected Satellite Communications," blog post, Intelsat, May 13, 2015. As of August 8, 2022:
https://www.intelsat.com/resources/blog/a-new-class-of-protected-satellite-communications/

Nightingale, Emily S., Bhavya Lal, Brian C. Weeden, Alyssa J. Picard, and Anita R. Eisenstadt, *Evaluating Options for Civil Space Situational Awareness (SSA)*, Institute for Defense Analyses Science and Technology Policy Institute, August 2016.

Northern Sky Research, *Satellite Ground Segment: Moving to the Cloud*, January 2021.

Northern Sky Research, "NSR's Satellite Capacity Report Sees Industry Moving Past COVID-19 Contraction to Drive $207B in Revenue Amidst Competition, Innovation and Risk-Taking," press release, July 27, 2022.

O'Callaghan, Jonathan, "The Risky Rush for Mega Constellations," *Scientific American*, October 31, 2019.

OneWeb, "OneWeb Confirms Successful Launch of 34 Satellites, Delivering Ongoing Momentum at the Start of 2022," press release, February 10, 2022a.

OneWeb, "Eutelsat and OneWeb to Combine: A Leap Forward in Satellite Connectivity," press release, July 25, 2022b.

Open Innovation Lab, "Standards and Specifications," webpage, undated. As of September 2, 2022:
https://apntoil.army.mil/standards-and-specifications

Orbit, "Orbit's GX30 Terminal Receives Type Approval to Operate on Inmarsat Global Xpress Commercial and Military Ka-Band Networks," press release, August 16, 2022.

Peterson, Brian, "Protected Tactical Waveform (PTW) Engineering Overview," Headquarters Air Force Space Command/A5M, March 16, 2016, Not available to the general public.

Pierre, Jeffrey, "After a Year of Deadly Weather, Cities Look to Private Forecasters to Save Lives," NPR, December 16, 2021.

Public Law 98-365, Land Remote-Sensing Commercialization Act of 1984, July 17, 1984.

Public Law 102-555, Land Remote Sensing Policy Act of 1992, October 28, 1992.

Rainbow, Jason, "Xona Space Systems Fully Funds GPS-Alternative Demo Mission," *SpaceNews*, September 22, 2021a.

Rainbow, Jason, "Tomorrow.io to Grow Weather Constellation Through SPAC Deal," *SpaceNews*, December 7, 2021b.

Rainbow, Jason, "Space X Requests Spectrum to Upgrade Starlink Mobile Services," *SpaceNews*, July 27, 2022a.

Rainbow, Jason, "Lockheed Invests in Xona's GPS-Alternative Constellation," *SpaceNews*, August 3, 2022b.

Raymond, John W., *Chief of Space Operations' Planning Guidance*, U.S. Space Force, November 2020.

Reece, Jared, "Infrastructure Asset Pre-Approval (IA-Pre) Kickoff," briefing slides for presentation to Commercial Space Cryptographic Cybersecurity Working Group, Space Systems Command, May 26, 2022.

Reksulak, Michael, William F. Shughart II, and Robert D. Tollison, "Innovation and the Opportunity Cost of Monopoly," *Managerial and Decision Economics*, Vol. 29, No. 8, December 2008.

Ryan, Gery W., and H. Russell Bernard, "Techniques to Identify Themes," *Field Methods*, Vol. 15, No. 1, February 2003.

Satelles, "Satelles and NIST Team Up for Precision Timing," March 30, 2022.

SES, *Consolidated Financial Statements as at and for the Year Ended 31 December 2019 and Independent Auditor's Report*, March 2, 2020.

SES, *Consolidated Financial Statements as at and for the Year Ended 31 December 2020 and Independent Auditor's Report*, February 25, 2021.

SES, *Consolidated Financial Statements as at and for the Year Ended 31 December 2021 and Independent Auditor's Report*, February 27, 2022.

Sheetz, Michael, "In Race to Provide Internet from Space, Companies Ask FCC for About 38,000 New Broadband Satellites," CNBC, November 5, 2021.

Sherman, Jeffrey A., Ladan Arissian, Roger C. Brown, Matthew J. Deutch, Elizabeth A. Donley, Vladislav Gerginov, Judah Levine, Glenn K. Nelson, Andrew N. Novick, Bijunath R. Patla, Thomas E. Parker, Benjamin K. Stuhl, Douglas D. Sutton, Jian Yao, William C. Yates, Victor Zhang, and Michael A. Lombardi, *A Resilient Architecture for the Realization and Distribution of Coordinated Universal Time to Critical Infrastructure Systems in the United States*, National Institute of Standards and Technology, Technical Note 2187, November 2021.

Siegel, Julia, "Commercial Satellites Are on the Front Lines of War Today. Here's What This Means for the Future of Warfare," Atlantic Council, August 30, 2022.

Space Capital, *Space Investment Quarterly: Q4 2022*, 2022.

Space Operations Command Operations Order 23-01, U.S. Space Command, December 2022.

Space Systems Command, "SSC CSCO Reaches Critical Milestone for IA-Pre, Roll-Out Begins Today," press release, May 26, 2022.

SPI—*See* U.S. Space Command Instruction.

Strout, Nathan, "Space Force Will Set Up One Office for Commercial Services, Including SATCOM and Satellite Imagery," C4ISRNET, June 2, 2021.

Swinhoe, Dan, "Data Centers with Dishes: How the Cloud Is Driving a Merger Between Data Centers and Ground Stations," DCD, December 22, 2021.

Telesat, *Annual Report Pursuant to Section 13 or 15(d) of the Securities Exchange Act of 1934 for the Fiscal Year Ended December 31, 2020*, U.S. Securities and Exchange Commission, commission file 333-159793-01, March 4, 2021.

Telesat, *Annual Report Pursuant to Section 13 or 15(d) of the Securities Exchange Act of 1934 for the Fiscal Year Ended December 31, 2021*, U.S. Securities and Exchange Commission, commission file 001-39240 (Telesat Corporation) and commission file 333-255518-01 (Telesat Partnership LP), March 18, 2022.

ThinKom PR, "SATCOM for U.S. DoD to Be Delivered by ThinKom, Inmarsat Government," press release, October 4, 2022.

U.S. Army, "Satellite Communications (SATCOM)," June 2, 2016.

U.S. Army Acquisition Support Center, "Assured-Positioning, Navigation and Timing (A-PNT)—Dismounted," webpage, undated. As of September 27, 2022: https://asc.army.mil/web/portfolio-item/a-pnt-dismounted/

U.S. Department of Defense, *Defense Space Strategy Summary*, June 2020.

U.S. Government Accountability Office, *Satellite Communications: DOD Should Develop a Plan for Implementing Its Recommendations on a Future Wideband Architecture*, GAO-20-80, December 2019.

U.S. Government Accountability Office, *GPS Modernization: DOD Continuing to Develop New Jam-Resistant Capability, but Widespread Use Remains Years Away*, GAO-21-145, January 2021a.

U.S. Government Accountability Office, *Satellite Communications: DOD Should Explore Options to Meet User Needs for Narrowband Capabilities*, GAO-21-105283, September 2021b.

U.S. Space Command Instruction 3250.01A, *Satellite Communications*, U.S. Space Command, May 18, 2020.

Viasat, *Annual Report Pursuant to Section 13 or 15(d) of the Securities Exchange Act of 1934 for the Fiscal Year Ended March 31, 2020*, U.S. Securities and Exchange Commission, commission file 000-21767, May 28, 2020.

Viasat, *Annual Report Pursuant to Section 13 or 15(d) of the Securities Exchange Act of 1934 for the Fiscal Year Ended March 31, 2022*, U.S. Securities and Exchange Commission, commission file 000-21767, May 27, 2022.

Wall, Mike, "SpaceX Launches 46 Starlink Satellites, Lands Rocket at Sea," Space.com, last updated August 31, 2022.

Weinzierl, Matthew, and Mehak Sarang, "The Commercial Space Age Is Here," *Harvard Business Review*, February 12, 2021.

Werner, Debra, "Startups Map Out Strategies to Augment or Backup GPS," *SpaceNews*, August 4, 2021.

White House, *National Space Policy of the United States of America*, December 9, 2020.

Wong, Jonathan P., Obaid Younossi, Christine Kistler LaCoste, Philip S. Anton, Alan J. Vick, Guy Weichenberg, and Thomas C. Whitmore, *Improving Defense Acquisition: Insights from Three Decades of RAND Research*, RAND Corporation, RR-A1670-1, 2022. As of September 6, 2022:
https://www.rand.org/pubs/research_reports/RRA1670-1.html

Xona Space Systems, "Xona Pulsar," webpage, undated. As of September 26, 2022:
https://www.xonaspace.com/pulsar

"Xona Space Systems Secures Million$$$ in Investments from First Spark Ventures + Lockheed Martin Ventures," Satnews, August 3, 2022.

Yan, Zhang, and Ryan Woo, "China's Geely Launches First Nine Low-Orbit Satellites for Autonomous Cars," Reuters, June 2, 2022.

Yin, Robert K., *Case Study Research: Design and Methods*, 5th ed., SAGE Publications, 2013.

Yonekura, Emmi, Brian Dolan, Moon Kim, Krista Romita Grocholski, Raza Khan, and Yool Kim, *Commercial Space Capabilities and Market Overview: The Relationship Between Commercial Space Developments and the U.S. Department of Defense*, RAND Corporation, RR-A578-2, 2022. As of August 25, 2022:
https://www.rand.org/pubs/research_reports/RRA578-2.html